OCR GCSE Specification B

Geography

Controlled Assessment
Student Workbook

David Payne

Consultants: John Belfield and Wayne Thomas

GW00420051

www.heinemann.co.uk

✓ Free online support
✓ Useful weblinks
✓ 24 hour online ordering

0845 630 33 33

Part of Pearson

Heinemann is an imprint of Pearson Education Limited, a company incorporated in England and Wales, having its registered office at Edinburgh Gate, Harlow, Essex, CM20 2JE. Registered company number: 872828

www.heinemann.co.uk

Heinemann is a registered trademark of Pearson Education Limited

Text © Pearson Education Limited 2010

First published 2010

12 11 10

10 9 8 7 6 5 4 3 2 1

British Library Cataloguing in Publication Data
A catalogue record for this book is available from the British Library.

ISBN 978 0 435 04378 0

Edited by Elina Helenius

Designed by Pearson Education Ltd

Typeset by AT Communication

Original illustrations © Pearson Education Ltd 2010

Picture research by Elena Wright

Cover photo © Rick Fischer / Masterfile

Printed in the UK by Ashford

Acknowledgements
The author and publisher would like to thank the following individuals and organisations for permission to reproduce photographs:

David Payne, p.56, 57, 71; Greg Balfour Evans/Alamy p.57.

The author and publisher would like to thank the following individual and organisation for permission to reproduce copyright material:

p.23 'Loved to death', adapted from *Loved not wisely but too well* by Carolyn Murrow-Brown,14 April 2001, *The Times*.

Every effort has been made to contact copyright holders of material reproduced in this book. Any omissions will be rectified in subsequent printings if notice is given to the publishers.

Websites
The websites used in this book were correct and up-to-date at the time of publication. It is essential for tutors to preview each website before using it in class so as to ensure that the URL is still accurate, relevant and appropriate. We suggest that tutors bookmark useful websites and consider enabling students to access them through the school/college intranet.

Contents page

Section 1: Getting the most out of your Controlled Assessment

What is Controlled Assessment?

Controlled Assessment is the method used to complete the part of your Geography GCSE called Geographical Enquiry (Unit B562). It is worth 25 per cent of your GCSE marks. The remaining 75 per cent of the marks are allocated to two written examinations. Your final GCSE grade will be determined by putting together the marks awarded for your **Geographical Enquiry** and the two written examinations.

What does the Geographical Enquiry unit involve?

The **geographical enquiry** unit involves completing the following two pieces of work during your GCSE Geography course of study.

1 **Fieldwork Focus** (15%) – You need to produce a fieldwork report based on an investigation carried out in a small scale area. This investigation should include the collection of **primary data**.

2 **Geographical Investigation** (10%) – You need to produce a report, PowerPoint presentation or poster. This will be based on the investigation of a particular topic or issue, using mostly **secondary data**. This can be at any scale.

The **Geographical Enquiry** unit is called Controlled Assessment because it has to be completed under supervision. This is because of its importance as part of your final GCSE grade.

What does Controlled Assessment mean?

All three parts of your GCSE Geography assessment (the Geographical Enquiry and two written examinations) are managed by the Examination Board (OCR). The Examination Board gives schools strict guidelines about how the geographical enquiry unit must be managed. These guidelines are called **'controls'** and they operate at three levels (limited, medium and high levels of control).

The table, on page 2, gives some guidance about how the control levels work.

MANAGING THE GEOGRAPHICAL ENQUIRY (UNIT B562)

1   TASK SETTING   High level of control

The Examination Board (OCR) set the task titles for both the Fieldwork Focus and the Geographical Investigation.

2   TASK TAKING – Completing the task   Limited level of control

a   Background research / planning / gathering Information and data
   – Some work can be completed without direct supervision.
   – Students can work together to gather information.
   – Teachers can offer help / guidance about data presentation.

b   Writing the report (applying data to the question)   High level of control
   – Carried out under direct, supervised conditions.
   – The report must be totally individual.
   – All work should be collected in after each session.
   – Limited teacher help (general guidance / clarification).
   – Research files and notes can be used – no new work can be brought in at this stage.

> **This is about:**
> – analysing data
> – reaching a conclusion
> evaluating your investigation.

3   TASK MARKING   Medium level of control
   – The Investigation is marked by your school using the Examination Board (OCR) mark scheme.
   – The marking is moderated (checked) by the Examination Board.

What do Controlled Assessment tasks actually test?

Controlled Assessment tasks test a number of things, including:

• the level of background knowledge you show about the topic investigated
• how well you identified the information needed to complete your investigation
• how effective you were at collecting information from a number of different sources
• how well you presented information through the use of a variety of techniques
• how well you identified the main points from your collected information
• how well you show that you understand the topic being investigated
• how well you show that you understand that information is not always reliable or detailed enough to make explanations complete.

What are the advantages of Controlled Assessment?

Although completing a geographical investigation can be quite challenging, it has many advantages, some of which are shown below.

> It is a part of your Geography GCSE that you have some control over.

> The fieldwork focus report might also come up in the final exams.

> It is worth 25 per cent of your GCSE, so could make a big difference to your final grade.

> It is a good way to develop your understanding of a particular part of your GCSE course.

> It gives you an opportunity to look in more detail at a topic you have enjoyed in class.

Safety First!

Completing a Geographical Enquiry involves collecting information outside of school, so always:

- listen to advice about safety
- discuss any individual data collection plans with teachers and parents
- do not go to places alone
- keep people informed about what you are doing.

Always think about **RISK ASSESSMENT** when carrying out work outside of school.

OVER TO YOU

Consider the potential risks in carrying out investigations in the following areas and suggest how the risks might be reduced.

Urban study in a busy town centre	River study in a remote rural area
Potential risks	Potential risks
How might risks be reduced?	How might risks be reduced?

Section 2: What you have to do

Introduction

The **Geographical Enquiry** part of your GCSE Geography course is made up of two parts, both of which will be completed at some time during the GCSE course. (They do not have to be completed at the same time).

The two parts are:

Fieldwork Focus – worth 15 per cent of the total GCSE marks

- OCR sets four fieldwork focus tasks each year. You have to investigate one of the tasks and produce a fieldwork report.
- The focus of this piece of work is **Fieldwork**. This means that your investigation:
  - should be based on a local (small scale) area
  - should be based on first hand information (primary data), although appropriate secondary data can also be used.
- It is recommended that the fieldwork report:
  - should be no more than 1200 words in length
  - should not take more than 10 hours class time (excluding background work and data collection).

Geographical Investigation – worth 10 per cent of the total GCSE marks

- The Examination Board (OCR) sets eighteen geographical investigation tasks each year, based on nine geographical themes. You have to complete one of the tasks and produce a report.
- The focus of this piece of work is **Investigation**.
  - It could be based on a topic in any location.
  - The chosen task could be at any scale.
  - It will have secondary data research focus, although appropriate primary data can also be used.
- It is recommended that the investigation report:
  - should be no more than 800 words in length
  - should not take more than 6 hours class time (excluding background work and data collection).

Word control

The total word count for the complete Geographical Enquiry (both pieces of work) should not exceed 2000 words – 1200 words for the Fieldwork Focus and 800 for the Geographical Investigation.

What is included in the word count?

The Examination Board states that:

- headings included in the presented work should be included in the word count
- figures, tables, diagrams, charts, footnotes and appendices should not be included in the word count.

What if my work is more than 2000 words?

The Examination Board states that:

- 'the completed work must not exceed 2000 words in total. If a candidate's work exceeds this, the final mark will reflect this'. This means that if your work is over 2000 words you may lose marks!

General guidelines

The following tips are taken from a 'General Guidelines' booklet produced by OCR.

Research, data collection and planning

Things to think about / remember…

- Make a plan of how you will spend the time you have for research / data collection. This way, you can make sure that you have time to cover everything you want to do.
- Make sure that you keep a record of where all the information you want to use comes from. This will allow you to include references and a bibliography when you write up the task.
- Think about how you will use your research or the data that you have collected to respond to the task. It may be helpful to make a basic plan so that you can check you have all the information that you need.
- Remember, you will not have access to resources other than your notes when you write up the task, so you need to make sure that you have all the information that you need in your notes.

During data collection, you can talk to your teacher about the task and ask them for advice. You can also work with other candidates and share ideas about the task with them. With out-of-classroom fieldwork, group data collection is allowed.

Writing up your reports

Things to think about / remember…

- Make sure that you include all the relevant information from your notes.
- Remember that it must be your own work.
- Remember that if you quote from another source (for example a book or the Internet) you must acknowledge this properly.

You will have access to all the notes that you made during the research / data collection period.

It will probably take several hours to write up your findings, but you will not have to do this all in one go. At the end of each session your teacher will collect in your work and your notes. They will give these back at the start of the next session.

Presenting your work

Things to think about / remember…

- The use of ICT is desirable. It could be used to:
  - carry out research on the internet
  - draw graphs and annotate photographs
  - word process reports.
- Maps, tables, graphs, spreadsheets, etc. should be inserted into the report at the appropriate place.
- Any copied information should be fully acknowledged.

> **Definitions**
>
> **Primary data** – Original data collected first hand by fieldwork (counting, measuring, asking questions, etc).
>
> **Secondary data** – Information from published sources (books, websites, newspapers, etc).

Fieldwork Focus

Selecting a Fieldwork Focus question

OCR sets four different 'Fieldwork Focus' questions each year. The four questions will be based on the following parts of your GCSE course.

- Rivers
- Coasts
- Population and settlement
- Economic development

You have to produce a report based on one of the Fieldwork Focus questions. The questions are published two years in advance and change each year. The following example shows the Fieldwork Focus questions for submission in 2011.

FIELDWORK FOCUS

Candidates must investigate one of the four fieldwork focus questions below.

Rivers
How successful are the flood management strategies for your chosen stream/river?

OR

Coasts
How and why do coastal features vary in your chosen area?

OR

Population and settlement
What are the issues that need to be considered if new housing were to be built in your chosen area?

OR

Economic development
How has an economic activity created conflict in your chosen area?

The tasks shown are for submission **in 2011 only**.

More information can be found on the OCR website (www.ocr.org.uk)

Do you have a free choice of Fieldwork Focus question?

There are two ways to approach the fieldwork focus question.

1 Each student selects one of the questions set by the examination board. You then:
- break the question down into smaller 'key questions'
- justify (explain) how these key questions will help you to answer the main question.

2 The school selects one of the questions set by the Examination Board. Then:

- the question is discussed in class or in small groups
- from the discussion key questions are identified
- you have to justify (explain) how these key questions will help you to answer the question.

What are 'key questions'?

Key questions are used to break down the original question set by the Examination Board into more manageable parts. They are often used to:

- put the question into a local context
- identify specific information that might be helpful
- help you to plan and justify your data collection methods.

The following example may help you think about how you can use key questions.

Fieldwork focus question – How can retail service provision in a town centre be improved?		
Possible key questions	**Justification**	**Data collection**
What is the existing retail service provision?	• Useful to find out the general range of shops / services.	Land use survey
What do people think about the general shopping quality?	• Useful to assess the views of local people about existing shopping quality. • Useful to find out what people feel needs to be improved.	Shopping quality matrix Questionnaire

What are people's shopping habits?	• Useful to find out the extent to which people are attracted to the area
	  – how often they visit
	  – where they come from
		Questionnaire

What is the general shopping environment like?	• It is not just shops that attract people. Other factors like parking, general facilities, pollution, litter, etc. may be important.	Photographs Environmental quality surveys
What could be done to improve the area?	• Find out how local planners and business could help to improve the area.	Interviews – local planners, business managers

Good Advice

Having drafted a list of key questions like this, you can then go back and adjust, remove some or add more.

Use the following space to start thinking about your own key questions.

OVER TO YOU

Fieldwork focus question

Possible key questions	Justification	Data collection

Good Advice

It is worth spending time putting together a set of really good key questions because they will then provide you with a clear plan of action.

Managing the Fieldwork Focus

The Fieldwork Focus is really about gathering enough information so that you can answer your chosen question. In order to do this the Examination Board identifies the following stages that you should go through. If you work through these stages it will give you a good opportunity to score the highest marks.

Four stages for managing the Fieldwork Focus

1 Setting the scene
- Choosing a question.
- Thinking about how the topic fits in with your geography course (what is it actually about?).
- Thinking about how the topic fits in with the study area.
- Identifying and justifing a number of key questions.
- Thinking about what the results of your data collection might be.
- Producing a location map of the study area.

2 Method of data collection
- Describing the methods used to collect primary data.
- Explaining why those methods were used to collect the data.
- Mentioning any problems or limitations associated with the data collection.

3 Data presentation and analysis
- Using a range of appropriate methods to present the data.
- Describing the key points identified from the collected data.
- Explaining the importance of the key points in relation to the original question.

4 Evaluation and conclusion
- Writing a substantiated conclusion which answers the key questions.
- Commenting on how well the investigation answered the original question.
- Commenting on any weaknesses / limitations of your investigation.
- Making suggestions about how your investigation could be improved / developed.

Good Advice

Use these points as a checklist as you work through your Fieldwork Focus investigation.

Making a start

There are three main parts to producing a successful fieldwork report.

1 Pre-fieldwork
Think about:
- the background to your investigation
- the actual area where your investigation is taking place
- key words and ideas
- planning the data collection methods.

2 Data collection
- Collecting primary data.
- Collecting secondary data.

3 Completing your report
- This is done in class under supervised conditions.

Pre-fieldwork

Locating your investigation

Is your investigation about a particular place, a part of a town or a general area? Identify the specific area you are investigating by looking at Ordnance Survey maps or town plans.

OVER TO YOU

Use this space to:

1 Define the actual area you are investigating.

2 Think about where you might find appropriate maps (location maps / base maps). This could include: planning office, estate agents, tourist offices, websites, etc.

Remember!

You can work with other people to collect primary data as long as it is clearly acknowledged and approved by your teacher.

Good Advice

Look carefully at the 'collecting and presenting information' section on pages 28–67 to give you some ideas.

Using Ordnance Survey (OS) maps

Ordnance Survey (OS) maps are available at different scales and can be used to locate the general area of a fieldwork investigation and to prepare detailed base maps of a local area.

Remember!

You have to produce a fieldwork report – not a book!

- You will have approximately 10 hours of class time to write up your work.
- Your finished report should be approximately 1200 words in length.
- Once you start writing up your work you will only have access to your notes (you cannot have access to any other resources). It is important that you put together a thorough set of background notes.

Key ideas / words

Every investigation has key ideas and words (often called 'geographical terminology'). The following example shows some key ideas and words that might be useful to an investigation about shopping patterns.

IDEAS

- People travel further to shop in larger shopping centres.
- Larger towns have more shops and services.
- People do not travel far for day-to-day goods.
  - Central business districts contain national stores.
    - Central shopping areas are often pedestrianised.

THINK

WORDS

Shops / services

Sphere of influence

Shopping hierarchy

Frequency of visit

Comparison goods

Convenience goods

Threshold population

Low / high order goods

Pedestrianisation

National chain stores

Good Idea

If there are a number of words that are very important to your investigation you could include a 'key definition' box in your introduction. You don't then have to keep explaining them throughout your report!

OVER TO YOU

Make a list of key ideas and words (geographical terminology) that might be useful to your investigation.

Key ideas	Key words

Background research

Background research is the first stage of any investigation. It is important because it will:

- *Provide valuable background information about the topic* – You might not end up using all of the information but it will help you to understand the context of your investigation (what it is about). It may also help you to decide what information is important for your investigation.
- *Identify any links there may be to geographical theory* – Some investigations are about comparing theory (what is expected) to an actual local example. To do this you need a good understanding about the original theory.
- *Help you to identify any helpful key words and definitions* – Using geographical words and definitions is a good way of identifying important ideas and showing understanding about the topic you are investigating.

Good Ideas

Keep all of your background information in a research file.

Make sure that you make a note of the sources of any information (including websites), because:

- you may need to find the sources again
- you need to include the sources of all information you have used in your bibliography.

Good Advice

You can get background information from textbooks, local newspapers, local authority websites and general websites.

The following example shows research notes for a river based investigation.

Fieldwork focus question – How and why do the features of a river vary? (The River Burr)

Basic river words / definitions that might be useful:

Source – where a river/stream begins

Tributary – a smaller stream flowing into a larger stream

Confluence – where streams / rivers join

Valley – a depression usually occupied by a stream (shape of a valley may change downstream)

Long profile – section of a river from the source ⟶ downstream

(Could use an OS map to draw this?)

Source

Burrbridge

Cross profile – section across the river (a number of cross-sections along the river might be useful)

Safety Factors

In this area the river is quite narrow and shallow.

The approximate length of this section of the river is 4km – it is very accessible.

Must consider

Risk assessment / safety!

The main river processes

Erosion: wearing away of rocks

Types of erosion

Hydraulic action – force of water

Abrasion – material carried by river acting like sandpaper

Attrition – material wearing away as it collides and rubs against itself

Solution – rocks dissolved by water

Bedload – material being carried by the river (could measure the type of bedload at different places · size of pebbles / roundess etc.)

Measuring the river

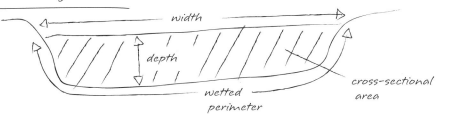

- measuring cross-sectional area (approximately) – width x depth
- measuring efficiency of river (hydraulic radius)   hydraulic radius = $\dfrac{\text{cross-sectional area}}{\text{wetted perimeter}}$
  (higher number = more efficient)
- measuring the speed of the river – time a float over a specific distance

Main feature of the River Burr is a meander (bend in the river)

Movement of material

Traction – rolling along river bed

Saltation – pebbles bouncing along river bed

Suspension – fine material (mud / silt) being carried by river

Solution – material dissolved in the water

Vertical – downwards erosion

Lateral – sideways erosion

What are the key ideas for the investigation?

1 How does the river valley change downstream?

2 How does the river change downstream?

3 Are there any specific features?

4 Does the bedload change downstream?

5 Does the river become more efficient as it flows downstream?

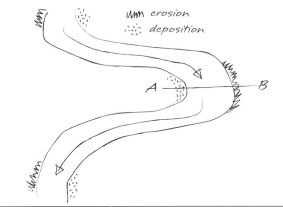

Use the following page to build up relevant background notes for your Fieldwork Focus investigation.

OVER TO YOU

Fieldwork Focus question ....

Research notes	Source of information

Planning the data collection methods

This is really about asking yourself two important questions:

- What do I need to find out?
- How am I going to find it out?

Think!

Always match up your key questions and your data collection methods.

OVER TO YOU

Use the 'collecting and presenting information' (pages 28–67) section to help you identify the data methods that would be helpful for your investigation. Always remember to justify your choice of data collection methods!

Data collection method	Justification

How will your fieldwork report be marked?

Your fieldwork report will be marked using an official mark scheme produced by the Examination Board (OCR).

The **mark scheme** identifies two areas for assessment (called **assessment criteria**). The following table describes the assessment criteria and briefly explains what they mean to you.

Assessment criteria	What does this mean?	Checklist
Application of knowledge and understanding	This means that you have to show that you can: • select and communicate knowledge and ideas • select and apply appropriate information to a particular topic/question • understand the relationship between people and environments • use geographical terminology accurately.	
Analysis and evaluation	This means that you have to show that you can: • collect and record information accurately • use a variety of skills and techniques to present and analyse data • use evidence to reach a conclusion • evaluate the validity of information.	

Good Idea

As you work through your investigation, use this marking criteria checklist to identify the strengths and weaknesses of your work by placing ticks or crosses next to each point in this column.

What is meant by 'analysis'?

Analysis is more than simply explaining the data. It means breaking the data down and picking out the most important points in relation to the original question.

How is the mark decided for each of the assessment criteria?

1 Each of the assessment criteria is divided into three levels, Level 1 being the lowest and Level 3 the highest.

2 The marks available for each level of the two assessment criteria are:

Assessment criteria	Level 1	Level 2	Level 3	TOTAL
Application of knowledge and understanding	0 – 4	5 – 8	9 – 12	12
Analysis and Evaluation	0 – 8	9 – 16	17 – 24	24
			Overall total	36

3 OCR produces a detailed mark scheme, which describes what is required to achieve the mark in each level.

What does the mark scheme look like?

The table below shows a simplified version of the mark scheme. The complete mark scheme can be found on the OCR website (www.ocr.org.uk).

Simplified mark scheme – Fieldwork Focus			
Assessment criteria	Level 1: Marks 0 – 4	Level 2: Marks 5 – 8	Level 3: Marks 9 – 12
Application of knowledge and understanding	• Simple descriptive background information about the topic. • Basic description and explanation of collected evidence.	• Applies background information to show a clear knowledge and understanding of the topic. • Shows some awareness of the context of the investigation. • Clear description and explanation of collected evidence.	• Applies background information to show a detailed knowledge and understanding of the topic. • Shows clear awareness of the context of the investigation. • Thorough description and explanation of collected evidence.
	Level 1: Marks 0 – 8	Level 2: Marks 9 – 16	Level 3: Marks 17 – 24
Analysis and evaluation	• Basic location and description of study area. • Some evidence of data collection from fieldwork. • Basic presentation using simple techniques. • Basic interpretation of information used to reach a simple conclusion. • Evaluation attempted. • Poor spelling, punctuation and grammar. • Poor organisational and communication skills. • Limited written work or written work largely not focused on the question.	• Uses some appropriate techniques to locate and describe the study area. • Range of evidence from different sources, including fieldwork. • Clear presentation using a range of maps, graphs, diagrams. • Information analysed and clearly used to reach a logical conclusion. • Clear evaluation which considers limitations of enquiry process and possible development. • Mostly accurate spelling, punctuation and grammar. • Clear organisational and communication skills. • Written work generally focused on the question and does not exceed the word limit.	• Uses a variety of techniques to locate and describe the study area. • Evidence from wide range of appropriate sources, including fieldwork. • Accurate presentation using a range of appropriate techniques. • Detailed analysis and interpretation of information to reach a thorough conclusion. • Detailed evaluation with limitations, possible solutions and suggestions about how the investigation could be extended. • Very few errors in spelling, punctuation and grammar. • Clear organisational and communication skills. • Written work well focused on the question and does not exceed the word limit.

OVER TO YOU

• Use the simplified mark scheme to help you understand what you are expected to do to achieve marks in the higher levels.

• Use highlighter pens to check off what you have done and identify what you need to do to reach Level 3 in each of the assessment criteria.

Geographical Investigation

Selecting a Geographical Investigation question

OCR sets 18 'geographical investigation' questions each year. The questions are based on nine geographical themes (two questions for each theme). You have to produce a report based on **one** of the Geographical Investigation questions. The questions are published two years in advance and change each year. The following example shows the Geographical Investigation questions for submission in 2011.

GEOGRAPHICAL INVESTIGATION

The geographical investigation offers a choice of nine themes.

Candidates must investigate one question from one of the themes.

Disease

1 How does the incidence of obesity vary across the UK?
OR
2 Is tuberculosis a major threat to the development of LEDCs?

Trade

3 What are the environmental impacts of the flower trade between the Netherlands and the UK?
OR
4 How global is a basket of vegetables in your local supermarket?

Ecosystems

5 How successful is the management of a sand dune ecosystem in the UK?
OR
6 How can we sustainably develop an area of tropical rainforest?

Sport

7 How does a chosen sports stadium bring advantages and disadvantages to its local area?
OR
8 How and why have the venues for Formula One racing changed over time?

Fashion

9 How can London Fashion Week bring advantages to the UK?
OR
10 What are the issues involved in "sweat shops"?

Energy

11 Should the UK develop nuclear power further in the future?
OR
12 How can a family in the USA reduce its energy consumption?

New technologies

13 How can new technologies help firms select a new location?
OR
14 How can new technologies help in monitoring pollution?

Crime

15 To what extent is gun crime an issue in Britain?
OR
16 What is the pattern of trafficking in people?

Tourism

17 How can the effects of visitors be managed in a national park you have studied?
OR
18 How sustainable are cruise holidays?

More information can be found on the ocr website (www.ocr.org.uk)

> The tasks shown are for submission **in 2011 only**.

Do you have a free choice of Geographical Investigation question?

You will have carried out a programme of study based on one of the nine geographical themes. Consequently, it is expected that your geographical investigation question will come from the theme that you have studied. For example, if you have studied Disease you will be expected to complete either question 1 or question 2.

The Geographical Investigation question can be selected by the school, or, if the school allows, students can choose which of the two questions to attempt.

Remember!

The Geographical Investigation requires **secondary data** research.

Primary data can also be used, if appropriate.

Managing the Geographical Investigation

The Geographical Investigation is about gathering enough information so that you can answer your chosen question. In order to do this the Examination Board suggests that your completed report should consider the following points:

1 – Location of the study / background information
2 – Breaking the question down into smaller investigation questions
3 – Issues that arise from the research
4 – Evaluation of research / data collection
5 – Conclusion (including looking at future possibilities).

What does this mean?

This means that you need to think about:

– the key ideas behind the theme you have studied
– the issues involved in the question you are looking at.

The following example shows how a Geographical Investigation question could be broken down in order to make it more manageable.

> **Examiner's advice**
>
> *What is an issue?*
> *An issue is where people have different opinions about a particular question. For example, when thinking about the question 'How should an area of tropical rainforest be developed?' there could be a lot of different ideas and some of these might conflict with each other.*

Theme: TOURISM

Question: *How important is tourism in a national park you have studied?*

Location / background

– *Where is the Lake District National Park (LDNP)?*
– *What are the characteristics of the LDNP? (population / settlements / industry / physical geography, etc.)*
– *How many people live in the LDNP?*

Investigation questions

– *How many people visit the LDNP each year?*
– *What is the pattern of visitor use?*
– *What proportion of local people rely on tourism for an income?*
– *To what extent are honeypot towns dominated by tourism?*
– *Is income from tourism used to improve social / environmental conditions?*

Issues identified

– *Is seasonality a problem?*
– *Are the effects of tourism throughout the LDNP evenly spread?*
– *Does tourism put pressure on the environment?*
– *Does tourism create conflicts?*
– *Could the LDNP survive without tourism?*

Evaluation of research

– *How reliable was the data collected?*
– *How objective / subjective are opinions about the issues?*

Conclusion

– *What are the advantages and disadvantages of tourism to the LDNP?*
– *How important is tourism to the local economy?*
– *How might the LDNP be managed in the future to ensure that tourism is sustainable?*

OVER TO YOU

Use the following space to break down your chosen investigation question and think about the issues involved.

Theme:	Question:

Remember!

This investigation is issue-based so you need to:

– show awareness of any issues involved

– show that different people have different opinions about particular issues.

Making a start

There are three main parts to producing a successful investigation report.

1 Background research

Thinking about:
- the locational context of your investigation
- background Information about the topic
- key words and ideas
- planning the data collection.

2 Gathering information

Collecting information relevant to the question.

3 Completing your report

This is done in class under supervised conditions.

Good Advice

Look carefully at the 'collecting and presenting information' section on pages 28–67 to give you some ideas.

Remember!

This will probably be mainly secondary research. This could include: facts and figures, case studies, photographs, different information about issues etc.

Good Advice

Make sure you identify and explain important issues linked to your chosen question.

What is an issue?

An issue is something that people have different opinions about or where there may be different ways of doing something.

For example – tourism in National Parks:

People who have tourism related jobs or businesses may be in favour of encouraging more visitors. Local people and environmentalists may think that increasing visitor numbers will create more noise, litter, congestion and damage the environment.

Remember!

You have to produce an investigation report – not a book!
- You will have approximately 6 hours of class time to write up your work.
- Your finished report should be approximately 800 words in length.

Examiner's advice

The key to a successful Geographical Investigation is planning. The question for investigation is really a discussion or debate where you have to make judgements about particular ideas.

Once you have chosen your investigation question ask yourself: 'What is this really about and what sorts of debate / arguments / differences of opinion will it raise?'

You really need to:
- *select and analyse a number of different sources of information so that you get a broad understanding of the question*
- *think about each source of information, for example,*
  - *How accurate is it?*
  - *Is it biased in any way?*
  - *Does it only give one side of an argument?*
  - *How well informed is it?*
  - *Is it factual or just opinions?*
- *weigh up all the evidence and use it to reach a conclusion that you can justify!*

Background research

The locational context of your investigation – your investigation may be about a place (a National Park) a country or even have a global context. Identify the specific area you are investigating by looking at maps.

OVER TO YOU
Use this space to:

1 Identify the locational context of your investigation.

2 Think about the different types of maps that might be useful for your Geographical Investigation and where you might find them.

Key words and ideas

Every investigation has key words (geographical terminology) and ideas. It is useful to identify these at the start of your investigation. You could include a 'definition box' in your introduction. You do not then have to keep explaining these words throughout your .

Remember!
Once you start writing up your work you will only have access to your notes so it is important that you put together a thorough set of background notes. (You cannot have access to any other resources.)

OVER TO YOU
Make a list of the key words (geographical terminology) and ideas that might be helpful to your investigation. (The example on page 19 may give you some guidance.)

Key words	Key ideas

Gathering background information

This is really the first stage of any investigation. It is important because it will:

- help you to understand the locational context and scale of the investigation
- help you to develop an understanding of the topic or area being investigated
- help you to identify any helpful key words and definitions.

You may not use all of your background research but it will increase your understanding about what your chosen question is about and it will help you decide what information is important.

Use the following two pages to build up relevant background notes for your Geographical Investigation. (The example on pages 12–13 may give you some guidance.

Good Advice

You can get background information from a variety of sources including: textbooks, atlases, newspapers, websites, etc.

Good Idea

- Keep all of your background information in a research file.
- Make sure that you make a note of the sources of any information (including web sites), because:
  - you may need to find the sources again
  - you should include a list of sources of information you have used in your bibliography.

OVER TO YOU

Investigation question ....

Research notes	Source of information

Research notes	Source of information

Planning the data collection

This is about asking yourself what precise information you need in order to answer the investigation questions that you have highlighted on page 00.

For example, looking at the example on page 00 about the Lake District National Park (LDNP) it might be important to find out:	
• The number of visitors to the LDNP and to different visitor attractions within the LDNP. • How many people live in the LDNP and what the age structure is like.	• The employment structure for the LDNP. • The extent to which honeypot towns are dominated by tourism facilities / businesses. • Conflicts / issues associated with tourism in the LDNP.
In order to find these things out there may be a number of places to look for information. For example: – National Parks in England and Wales (website) – Lake District National Park (Visitor Centre website) – Lake District Tourist Agency (website) – Individual tourist offices within LDNP – Local authority websites – Business directories for the Lake District – National Census data – Local newspapers – Friends of the Lake District (website)	

Remember!

This is a research task where the focus will be on the use of secondary data.

OVER TO YOU

Investigation question

What precise data do you need to find out?	Useful sources of information

Examiner's advice

You could keep a research diary where you log your sources of information. The following example will be useful when you write your investigation report.

Source of information	Summary of content	Why it was useful?	What were its limitations?

How will your Geographical Investigation be marked?

Your Geographical Investigation will be marked using an official mark scheme produced by OCR.

The **mark scheme** identifies two areas for assessment (called **assessment criteria**). The following table describes the assessment criteria and briefly explains what they mean to you.

Assessment criteria	What does this mean?	Checklist
Application of knowledge and understanding	This means that you have to show that you can: • use research to show that you understand geographical ideas • use your knowledge to explain that people have different opinions about issues • use information to reach conclusions.	
Analysis and evaluation	This means that you have to show that you can: • collect and record appropriate information • describe, explain and analyse evidence (data) • use evidence to reach a conclusion • evaluate the validity of information.	

Good Idea

As you work through your investigation, use this marking criteria checklist to identify the strengths and weaknesses of your work by placing ticks or crosses next to each point in this column.

What is meant by 'analysis'?

Analysis is more than simply explaining the data. It means breaking the data down and picking out the most important points in relation to the original question.

How is the mark decided for each of the assessment criteria?

1 Each of the assessment criteria is divided into three levels, Level 1 being the lowest and Level 3 the highest.

2 The marks available for each level of the two assessment criteria are:

Assessment criteria	Level 1	Level 2	Level 3	TOTAL
Application of knowledge and understanding	0 – 4	5 – 8	9 – 12	12
Analysis and evaluation	0 – 8	5 – 8	9 – 12	12
			Overall total	**24**

3 OCR produces a detailed mark scheme, which describes what is required to achieve the mark in each level.

What does the mark scheme look like?

The table below shows a simplified version of the mark scheme. The complete mark scheme can be found on the OCR website (www.ocr.org.uk).

Simplified mark scheme – Geographical Investigaton			
Assessment criteria	Level 1: Marks 0 – 4	Level 2: Marks 5 – 8	Level 3: Marks 9 – 12
Application of knowledge and understanding	• Basic application of knowledge to reach a conclusion. • Some awareness that people might have different views about an issue. • Basic use of evidence to support argument.	• Applies some geographical ideas to reach a valid conclusion. • Shows awareness of different attitudes and reactions to issues. • Research evidence clearly used to support argument.	• Applies a range of geographical ideas to reach a detailed conclusion. • Explains why people have different attitudes and reactions to issues. • Uses research evidence to support and justify argument.
	Level 1: Marks 0 – 4	Level 2: Marks 5 – 8	Level 3: Marks 9 – 12
Analysis and evaluation	• Limited amount of appropriate information used. • Little or no attempt to acknowledge sources of information. • Basic interpretation of information used to reach a simple conclusion. • Vague appreciation of the limitations of evidence. • Some attempt to organise report. • Poor spelling, punctuation and grammar. • Limited written work OR written work largely not focused on the question.	• A variety of appropriate information used. • Most sources of information acknowledged. • Information analysed and clearly used to reach a logical conclusion. • Recognises some limitations of the evidence used. • Clearly organised and logical report. • Spelling, punctuation and grammar mostly accurate. • Written work generally focused on the question and does not exceed the word limit.	• A wide range of clearly relevant sources used. • All sources of information acknowledged. • Detailed analysis and interpretation of information to reach a thorough conclusion. • Critical appreciation of all the evidence used. • Well structured and logical organisation of report. • Spelling, punctuation and grammar has very few errors. • Written work well focused on the question and does not exceed the word limit.

OVER TO YOU

- Use the simplified mark scheme to help you understand what you are expected to do to achieve marks in the higher levels.
- Use highlighter pens to check off what you have done and identify what you need to do to reach Level 3 in each of the assessment criteria.

Section 3: Collecting and presenting information

Introduction

Collecting the information you need in order to complete your reports is probably the most important part of the whole process. Without a good range of information it will be difficult for your work to score high marks. The diagram below explains why.

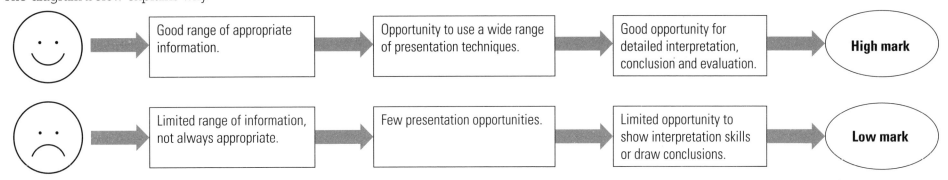

Primary and secondary data

There are two main types of information – usually called data – and these are shown in the table below.

Good Idea

When collecting primary data always make a note of when and where it was collected. Also make a note of any problems you may have had, and any limitations to the data collection methods that you used.

Type of data	Definition	Examples
Primary data	Original information collected first-hand by fieldwork.	Measuring Counting Assessing Interviewing Sketch mapping Taking photographs
Secondary data	Information from published sources which was collected by someone else.	Census data Textbooks Planning documents Local newspapers Leaflets Maps Directories Websites Photographs

Remember!

- Your Fieldwork Focus investigation must be based on primary data evidence, although secondary data can also be used.
- Your Geographical Investigation will be based on secondary data evidence, although primary data can also be used if appropriate.

Where should you start?

In order to get the highest marks for your reports, the information and data you collect must be clearly linked to the original question being investigated. Asking yourself the following two questions is a good place to start:

1 What information do I need to answer my question?

2 Why is this information important to my investigation?

Start with a 'thinking exercise' like the one below. This will help you to identify types of data that might be helpful.

Fieldwork Focus question	Should a new industrial development be built in your chosen area?
What information might be helpful?	Why might the information be helpful?
– Land use survey	– To identify the land uses in the area and assess whether industry would fit in.
– Survey of local services	– To see if there are enough local services to cope with added business.
– Traffic survey	– To see if the area already has traffic pressures.
– Questionnaire	– To get people's views about the issue.
– Environmental / quality survey	– To assess the existing quality of the area and consider whether a new industrial development would improve it.
– Interview – commercial estate agent	– To see if there is a demand for new industrial premises and if this is a good location.

OVER TO YOU

Use this space to do a 'thinking exercise' like the one on this page. (Include data that your teacher may have supplied.)

FIELDWORK FOCUS – Title

What information might be helpful?	Why might the information be helpful?

GEOGRAPHICAL INVESTIGATION – Title

What information might be helpful?	Why might the information be helpful?

Using this section to develop your investigations

It is important that the data you use for your investigations is clearly relevant to your chosen questions.

Use this section to:

- identify potential data collection methods that could be useful in addressing your chosen question
- identify different presentation techniques that you might use.

Fieldwork Focus

Remember that this is a fieldwork investigation so it must include primary evidence, but that appropriate secondary evidence can also be used.

Key points from the mark scheme

For the highest marks you have to:

- apply background information to show a clear knowledge and understanding of the topic
- use evidence from a wide range of appropriate sources, including fieldwork
- use a range of appropriate presentation techniques.

OVER TO YOU

Theme: .... **Question:** ....

Look through the whole of this section (including the secondary research methods) and identify any data collection and presentation methods that might be useful to your Fieldwork Focus investigation. Explain why each might be useful.

Data collection method	Page	Data presentation method	Page

Geographical Investigation

Remember that this is a research investigation so it involves detailed use of secondary evidence, but that appropriate primary evidence can also be used.

How could **primary** evidence be used?

- If the investigation question is based in the UK, questionnaires and interviews might provide useful evidence.
- Questionnaires and interviews are a useful way of identifying what people think about an issue and judging the strength of different opinions.

Key points from the mark scheme

For the highest marks you have to:

- use a wide range of relevant sources of information
- use research evidence to support your argument
- explain why people have different views and opinions about an issue.

OVER TO YOU

Theme: .... **Question:** ....

Look through the whole of this section (including the primary research methods) and identify any data collection and presentation methods that might be useful to your investigation. Explain why each might be useful.

Data collection method	Page	Data presentation method	Page

Primary data: using questionnaires

Questionnaires are an excellent source of primary data and can be used to obtain information about people's habits, people's views and opinions.

- Questionnaires are frequently used in Fieldwork Focus investigations. However, they can also be used in research investigations, especially where information about people's views and opinions are requested.
- Questionnaires are more frequently used in human geography investigations. However, it might be appropriate to use a questionnaire in a mainly physical geography investigation; for example, when considering the recreational use or management of a river or coastal area.

Constructing a questionnaire

A questionnaire must be carefully planned if it is going to provide useful data. It is not just about asking questions, it is about asking appropriate questions!

In order to produce an effective questionnaire you need to consider the following questions.

- What information will be useful to your investigation?
- How does each question relate to your investigation?

What types of information can you obtain from respondents (people who answer questionnaires)?

- Background data about the respondents, for example:
  - age and gender information
  - occupational data.
- Activity data, for example:
  - how often people go shopping
  - how often people visit a park
  - what local facilities people use.
- Attitude data, what people think, for example:
  - do you think a new road should be built?
  - what do you think about the environmental or shopping quality of an area?
  - do you think a particular place needs more facilities?

Types of questions

The types of question you ask will be determined by whether you need lots of basic information or a limited amount of more detailed information. The two main types of question are:

Closed questions

These are short-answer questions, often with a yes / no tick box response. They are an excellent way of collecting lots of information very quickly but do not always give detailed responses.

Trying out your questionnaire

It is always a good idea to test or 'pilot' your questionnaire on a small number of people. You can then adjust any questions that do not seem to work.

Remember!

Questions can give objective information (facts) and subjective information (opinions).

Open questions

These are questions where respondents can make longer, general comments. They can provide considerable detail but can take a long time, and answers may be more subjective (opinions).

How long should a questionnaire be?

The length of a questionnaire will depend upon what you are trying to find out, but usually a questionnaire should not have more than ten questions. Questionnaires with mainly open questions should have fewer questions.

How should you lay out a questionnaire?

The most successful questionnaires usually start with several closed questions which gather general information and end with a limited number of open questions.

How can you get personal data?

– When asking questions, note the sex and approximate age of the respondents (young / middle aged / older).

– Make questions less personal by offering a broad choice of options. For example:

'How old are you?' Tick one box.

☐ 15 – 25 ☐ 26 – 40 ☐ 41 – 55 ☐ 56 – 65 ☐ Over 65

How many people should you ask?

You are going to draw conclusions from the people you ask, so you need to get enough evidence to be sure that your results are a true representation of the general population (a 'representative sample'). For a shopping survey this might mean between 50 and 100 people; for a small village study, 10 to 20 people might be a reasonable representative sample. The number of people you ask might also be determined by the type of questions. If your questionnaire has only short questions, you could ask a larger number of people.

How do you select the people to ask?

Selecting people to answer a questionnaire is called sampling. You cannot ask everybody, so you need to select or sample a number of people.

– The type of people you ask must reflect the aim of your investigation. For example, in a study about car parking problems asking non-drivers or very young people will be of limited use.

There are three main types of sampling:

– **Random sampling** – where each member of a population has an equal chance of being selected. For example, in a street questionnaire every house number is put in a bag and ten are selected at random.

– **Stratified sampling** – where the proportion of respondents is selected according to the topic. For example, in a questionnaire about youth club facilities it might make more sense to ask a larger proportion of younger people.

Good Idea

Use a round number of questionnaires (10, 20, 50, 100 etc.). This will make it easier for you to work out percentages and pie charts.

Good Idea

When you write up your investigation explain:

- how you piloted your questionnaire
- how you decided the number and type of people to ask.

- **Systematic sampling** – where a regular sample is taken. For example, every tenth person, every fifth house, every ten metres.

Safety first

When planning questionnaires always make sure your parents and teachers know what you are doing, and always work in pairs.

OVER TO YOU

Fieldwork focus question	**Geographical investigation question**
1 List the types of information it would be helpful to find out from a questionnaire.	1 List the types of information it would be helpful to find out from a questionnaire.
2 Think about the number and type of people you might ask.	2 Think about the number and type of people you might ask.
3 Draft a number of questions that might help you get the information you need.	3 Draft a number of questions that might help you get the information you need.

Presenting questionnaire data

The methods you choose to present your questionnaire data will be determined by the types of questions you have asked.
The following techniques are frequently used to present information obtained from questionnaires:

- bar graphs (page 63)
- pie charts (page 64)
- tables of figures
- pictograms
- desire line maps.

Pictograms

Pictograms are picture graphs where the picture gives a visual representation of the data. For example:

Question: How did you travel here today?

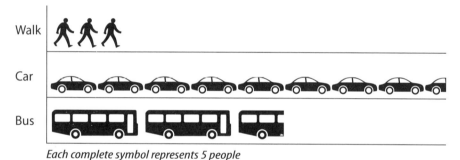

Each complete symbol represents 5 people

Desire line map

A desire line map is drawn to show the movement of people, either individually or in groups. For example:

Question: Which town/village did you come from today?

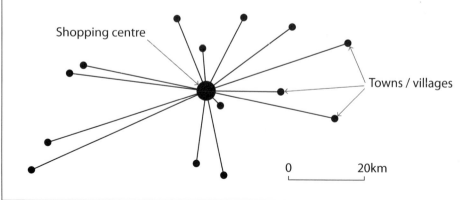

Primary data: using interviews

Interviews are a useful source of primary data. They can give you a lot of detailed and well informed information and can be more objective than a questionnaire.

Good Idea

Interviews can be used to support questionnaire data or get extra information.

Preparing for an interview

1 Think about the types of interviewees who could provide helpful information for your investigation.

2 (a) Work out exactly what you need to find out.

(b) Prepare a number of key questions that will help you to focus on the information you need.

3 Think about any additional secondary data that the interviewee might have that could be useful to your investigation.

4 Always remember that interviewees are helping you and giving up their time, so create a good impression by being well presented, well prepared. Aim to be punctual and well organised and do not take too long.

For example:

For this investigation you could interview the people below.

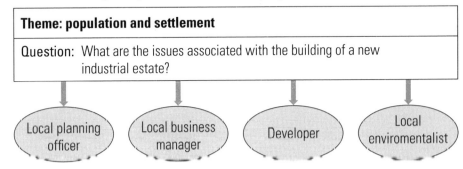

Theme: population and settlement

Question: What are the issues associated with the building of a new industrial estate?

Local planning officer · Local business manager · Developer · Local enviromentalist

Presenting interview information

A ten-minute interview may run into hundreds of words! When you go back over an interview the key is to identify the most important points. (Those that are closely linked to the title of your investigation.) These points need to be highlighted when you present your interview data.

How can the key points be presented?

There are a number of ways that interview data can be presented, including:

1 a simple list of the most important points

2 a list of the key points mentioned by each interviewee. (Important when there are different opinions about something.)

3 using speech bubbles and writing a brief statement identifying the key points mentioned by each interviewee.

The advantage of using speech bubbles is that it clearly identifies the views of different people and is a visual presentation.

However – you must make sure that the statement in each bubble is an accurate representation of the points made by the interviewee!

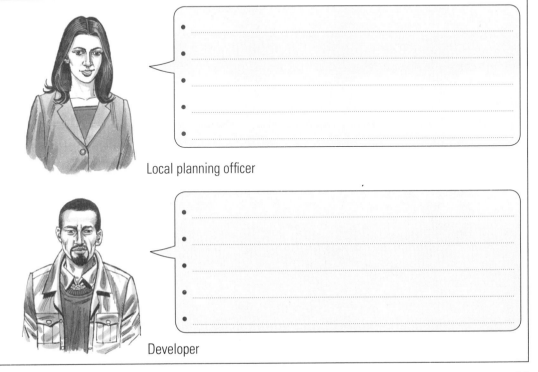

Local planning officer

Developer

OVER TO YOU

Use the following space to work out which people might provide useful information for your investigation and what questions you might ask.

Fieldwork Focus investigation

Geographical Investigation

Good Ideas

- If you are not sure about your questions try them out on a friend first. If they are not clear or appear to be too personal make adjustments to them.
- You could tape interviews but always ask the interviewee if they mind!

Safety First

- Discuss any individual interviews you may be thinking of doing with your teacher or parents.
- Work in pairs – one person asking questions while the other takes notes.

Primary data: using simple scoring surveys

A simple, graded scoring survey is a good way of gathering information and can be used in a variety of investigations.

Basic number / description grids like those shown below will often give a good first impression of what you are trying to find out. Completing a number of graded scoring surveys in an area is a useful way of identifying differences and making comparisons.

The following examples show simple scoring surveys which could be used to collect different types of data.

You can use this technique and construct your own scoring system to fit your investigation.

Noise pollution	
Noise level	Description
1	Can hear a whisper
2	Can hold a normal conversation
3	Have to raise voice to be heard
4	Difficulty hearing conversation

Footpath erosion	
Level of damage	Description
1	No evidence of damage
2	Some evidence of damage
3	Generally worn
4	Serious damage / hazardous

Traffic congestion	
Traffic movement	Description
1	Traffic moving freely
2	Minor hold-ups at junctions
3	General congestion
4	Considerable congestion

Water pollution	
Water quality	Description
1	No evidence of pollution
2	Some evidence of pollution
3	Generally polluted
4	Highly polluted

Good Idea

You can explain the descriptors in your grid in more detail. For example, in the water pollution grid:

'No evidence of pollution', means that the water is clear and there are a lot of plants and animals evident whereas 'Highly polluted' means that the water is totally cloudy with surface pollution and no evidence of plant and animal life.'

What are the advantages of using simple scoring surveys?

- A lot of information can be gathered quickly, without the use of equipment.
- It can give a useful first impression and can then be developed further if the information appears to be important.
- While it might not always be easy to identify the difference between one number and the next (1 and 2, etc.), it should be easy to see the difference between the extremes in the grid.

Good Advice

Do not forget to explain the advantages and disadvantages of your data collection methods in your report!

What are the limitations of using simple scoring surveys?

- They are not very scientific and can lack accuracy.
- It may not be easy to judge between levels.
- Judgements are subjective – different people may have different opinions!

Presenting scoring survey data

Simple scoring survey data can be presented in a table of figures. However, it is much better to use a map where you can show exactly where it was collected.

A proportional symbol map like the one shown here is a useful technique.

Water pollution

Farm

Farm

N

Highly polluted

No evidence of pollution

Town

Heavy industry

Direction of flow

0 1 2 3 4 5km

OVER TO YOU:
For your investigation

List the types of information that you could collect using a simple scoring survey.

Complete the following data collection grid for one type of information that you could collect using a simple scoring survey.

Heading:

	Description
1	
2	
3	
4	

Primary data: using more detailed scoring surveys

Simple scoring surveys (pages 38–39) are a quick way of getting basic information from a number of places. Sometimes this may be all that is needed for an investigation. However, if more detailed or specific information is needed a complex scoring / ticking system might be useful.

The following examples show two different types of scoring / ticking systems, each of which could be modified to suit different investigations.

An amenity index - used to assess the functions in different settlements.

Function	Settlement A	Settlement B	Settlement C	Settlement D
Post office	✓		✓	✓
Supermarket			✓	✓
Clothes shop				
Hairdressers			✓	✓
Newsagents	✓	✓	✓	✓
Travel agent				
Bank				✓
Cinema				
Police station				✓
Bus / train station				✓
Doctors			✓	✓
Dentist				✓
Primary school	✓		✓	✓
Secondary school				✓

The list of functions could be changed to suit the investigation.

Presentation

The total number of functions in each settlement could be added up and the totals presented using a proportional symbol map (page 39).

What are the advantages of using more detailed scoring surveys?
- They can give a more accurate and complete picture.
- They can include a wider range of information.
- They can include more objective data.

Ticks or numbers could be used (i.e. if a settlement has two supermarkets a number 2 would be put in the column).

Ticks show whether a settlement has the function or not.

The total number of functions in each settlement can then be calculated.

Residential quality index

Used to assess the quality of different residential areas in a town.

	1	2	3	4
	Very poor	Poor	Average	Good
Quality of decoration (paintwork, etc.)				
General maintenance (windows / fences, etc.)				
Garden (tidiness)				
Quality of open spaces (green areas / play areas)				
Traffic safety (volume of cars / parking, etc.).				

A total residential quality score can be calculated by adding up the five individual scores for each area. The lowest possible residential quality score would be 5, the highest would be 20.

A residential quality index is a useful way of getting sensitive information and can assess a range of factors. However the information can be rather subjective unless specific factors are included.

Presentation

Total residential quality scores could be presented using a proportional symbol map (see page 39).

Good Idea

The use of annotated photographs is helpful when comparing different areas.

A small number of photographs could be included on a proportional symbol map.

OVER TO YOU
Investigation question

1 List the types of information that you could collect using a detailed scoring / ticking system:

2 Explain why any two types of information that you have listed would be useful for your investigation:

1

2

Primary data: assessing environmental quality

Environmental quality assessment is not just about the countryside! It can be used to make judgements about a number of factors, including:

- the impact of river and coastal management schemes on the environment
- the quality of different parts of an urban area
- the suitability of housing or industrial development in a particular area
- how different types of economic activity affect environmental quality
- the quality of different retail areas within an area.

One of the easiest ways of visually assessing environmental quality is by using a technique called 'bi-polar analysis' (bi-polar means 'two poles' or opposites).

How do you use bi-polar analysis?

A Think about what it is you want to assess.

B Work out the positive and negative aspects of each category you use to make your assessment.

For example, if you want to compare the environmental quality of different parts of a town centre, the bi-polar grid below might be useful.

Positive pole	+3	+2	+1	0	-1	-2	-3	Negative pole
Low level of traffic noise		✓						High level of traffic noise
No litter					✓			Lots of litter
Attractive buildings		✓						Unattractive buildings
Well maintained	✓							Poorly maintained
Landscaping / seating		✓						No landscaping / seating
	Total score = +8							

C Complete the bi-polar grid by making a visual judgement and ticking the appropriate boxes. (The stronger the impression the higher the number.) Zero suggests that there is no strong feeling one way or the other. In the example above, the area being assessed scored a total of +8, suggesting that the environmental quality is average. However, it is also clear that the area has a slight litter problem.

D You can analyse bi-polar data by:
- looking at the total grid score for a place
- comparing total grid scores from a number of different places
- looking at the scores for individual categories within the bi-polar grid.

Remember!

Environmental quality data is often subjective (based on opinion rather than fact). This might be considered a weakness in its reliability as a source of data. You should mention this in your evaluation.

Good Idea

It is helpful to use a map to show where your bi-polar data was collected.

Presenting enviromental quality data

1 A proportional symbol map (page 39) could be used to compare different bi-polar scores in an area.

2 Bi-polar graphs can be used to:

a compare different places (in this case places A-E)

b compare the views of different people (here each line represents one person)

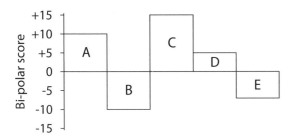

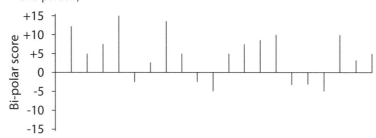

3 Using photographs with your bi-polar survey data is a good way of linking the bi-polar score with the actual area being assessed. It also means that the positive and negative points can be identified using annotations.

OVER TO YOU

Investigation question

1 Complete the bi-polar grid below by identifying the categories that could be used to measure one aspect of the environment in your investigation.

← *Write in this box what you are actually assessing.*

Positive pole	-3	-2	-1	0	+1	+2	+3	Negative pole
Total score								

2 Explain why this technique is useful to your investigation.

....

....

Primary data: using pedestrian surveys

Many investigations include an understanding about:

- how people behave
- how people interact with a settlement (town or village) or the natural environment.

Pedestrian surveys

Pedestrian surveys measure human behaviour and can be a very useful source of primary data. They can be used in a number of different types of investigation, some of which are shown by the diagrams below.

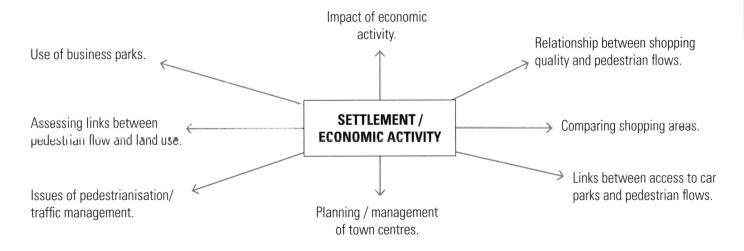

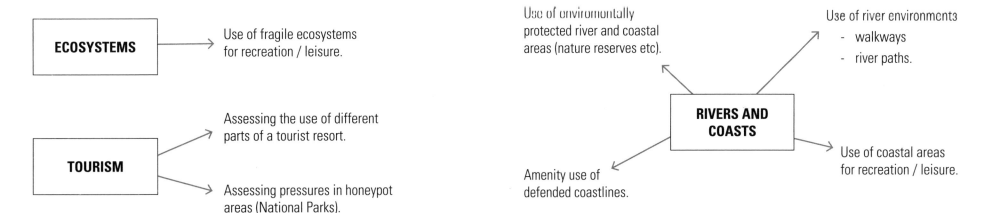

Pedestrian surveys might also be useful when:

a investigating the use of recreational areas such as beaches, riverside walkways, footpaths

b investigating the use of environmentally protected areas such as nature reserves, forest parks or other protected ecosystems.

Presenting pedestrian survey data

Two different ways of presenting pedestrian survey data are shown below.

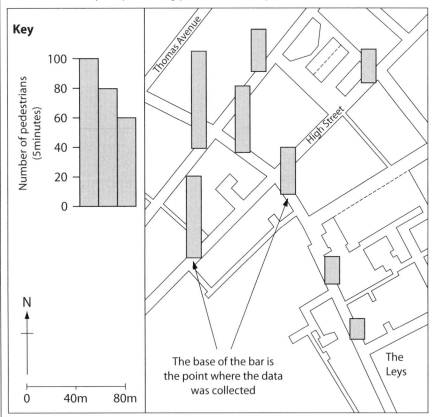

Key

Number of pedestrians (5 minutes)

The base of the bar is the point where the data was collected

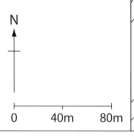

The Leys

Key

1. Plot your pedestrian flow numbers on a map.

2. Join up the same (or approximate) numbers as isolines.

3. Shade in your map to construct a choropleth map: the higher the number, the darker the shading.

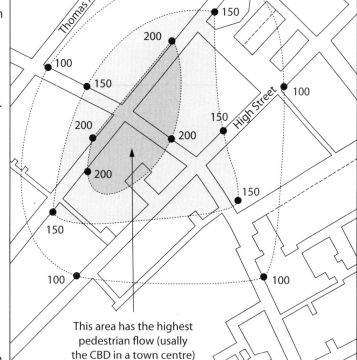

This area has the highest pedestrian flow (usally the CBD in a town centre)

To show the direction of pedestrian movement a flow line map could be used (see page 46).

Primary data: using vehicle surveys

Traffic flow data and parking surveys can be a useful source of primary data in many different types of investigation. The following diagram shows ways that vehicle surveys can be used:

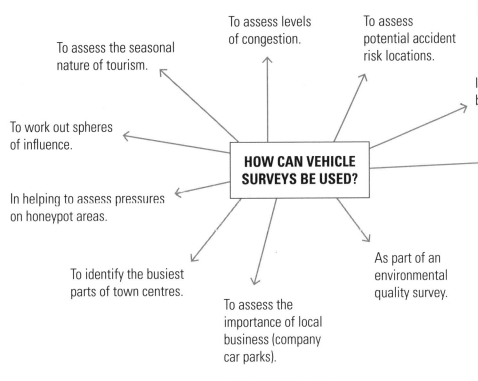

To assess levels of congestion.

To assess potential accident risk locations.

To assess the seasonal nature of tourism.

In helping to identify the central business district of towns.

To work out spheres of influence.

HOW CAN VEHICLE SURVEYS BE USED?

To assess parking problems.

In helping to assess pressures on honeypot areas.

To identify the busiest parts of town centres.

As part of an environmental quality survey.

To assess the importance of local business (company car parks).

Safety First

- Always choose safe places to collect traffic data.
- Work in pairs when collecting traffic and parking data.

Good Ideas

- If you need to compare traffic data make sure you use the same survey sites and times on different days.
- When completing traffic surveys make a note of:
  - the general conditions (weather)
  - levels of congestion.

  These factors might affect your results:
- Remember – traffic congestion adds to air pollution so could be an important environmental quality factor.

Presenting vehicle survey data

Flow line maps are a good way of showing traffic flows. The width of the line is drawn in proportion to the number of vehicles and an arrow shows the direction of flow.

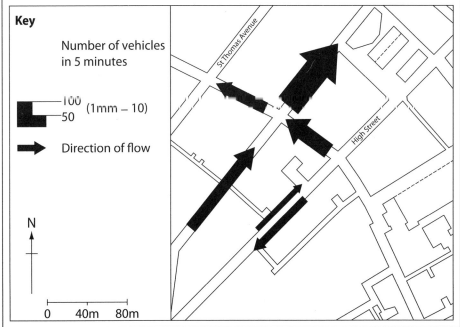

Key

Number of vehicles in 5 minutes

100
50
(1mm – 10)

→ Direction of flow

N

0   40m   80m

Traffic flow surveys

Traffic flow surveys record the total flow of traffic or the different types of traffic using particular roads. Constructing a data collection sheet, like the one below, will make data collection easier and more accurate.

Date		Time		
Location		Weather		
Cars	Motorcycles	Lorries / vans	Buses / coaches	Bicycles
ɪɪɪɪ ɪɪɪɪ ɪɪɪɪ ɪɪɪɪ ɪɪɪɪ ɪɪ	ɪɪɪɪ ɪɪ	ɪɪɪɪ ɪɪɪɪ ɪɪɪɪ ɪɪ	ɪɪɪ	ɪɪɪɪ ɪ

Car parking surveys

Car parking surveys are frequently used as a source of data in urban investigations. However, they are also useful in rural, tourism and honeypot area investigations. They can give a useful indication to:

1 parking problems

2 congestion / overcrowding issues

3 use of shopping areas / shopping quality

4 effectiveness of parking management.

Presenting car parking data

A proportional symbol map could be used to present car parking data (page 39).

Tax disc survey

Vehicle tax discs have written on them the name of the place where they were purchased. By looking at tax discs you can identify where vehicles may have come from. This could help you to identify the sphere of influence of areas, tourist areas or large shopping centres. (Ignore the tax discs issued centrally by the DVLA!)

Presenting tax disc information

A car park tax disc survey was carried out on twenty cars at Castleton – in the Peak District National Park. This desire line map was produced using the data collected by the survey.

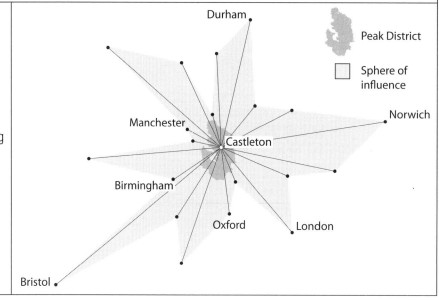

Sphere of influence

The area over which a place has influence (attracts people).

Safety First

Be careful, some people do not like their cars being looked at too closely!

OVER TO YOU
Investigation question

1 How might **pedestrian surveys** be useful for your investigation?

2 Use the space here to identify the places / times you might collect the data (use a rough sketch map of your study area).

OVER TO YOU
Investigation question

1 How might **vehicle flow surveys**, car parking surveys or tax disc surveys might be useful for your investigation?

2 Use the space here to identify the places / times you might collect the data (use a rough sketch map of your study area).

Primary data: assessing land use

A land use survey is a way of assessing the way that land is used in an area and can be either primary or secondary data.

Primary data - if you complete your own land use survey.

Secondary data - if you obtain a land use map / land use information from a secondary source (local authority planning department, estate agent, Ordnance Survey or a map, etc.).

Good Idea

You can often get useful base / street maps from the local authority planning office, Ordnance Survey or local estate agents.

Land use surveys can provide useful information for different types of geographical fieldwork investigation, including:

Settlement investigations:

- look at planning and issues such as the management of traffic and people
- compare the land use of a local area with urban models
- look at shopping quality and patterns of shops
- investigate land use change / development.

Economic development investigations:

- look at planning issues in relation to present and future locations of economic activities
- consider / compare the locations of economic activities
- consider industrial location factors
- assess the importance of economic activities in an area
- consider environmental issues in relation to economic activities.

Tourism investigations:

- looking at the influence/importance of tourism in an area.
- considering how the range and type of tourist facility might attract visitors.
- considering the balance between functions for local people and visitors in an area

There are three main types of land use survey:

General land use survey

This identifies the general land use in an area (commercial, residential, industrial etc.). It is not particularly detailed but gives a useful impression of the overall pattern of land use and the relative importance of different functions.

The example (right) identifies and shows the location of the dominant land uses in a tourist resort.

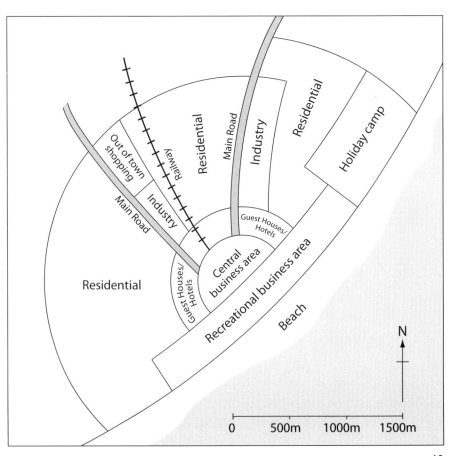

Detailed land use survey

This type of land use survey identifies the use of particular buildings or types of shops and services (usually on the ground floor). This type of land use survey is helpful when completing investigations about shopping habits, identifying central business districts (CBD) or topics linked to the management of town centres.

Carrying out a detailed land use survey

In small settlements it might be possible to identify every building. However, in urban areas this will not be possible. Instaed, identify the main categories of shop/service (especially the ones that are important to your investigation). The following example shows a detailed town centre land use map:

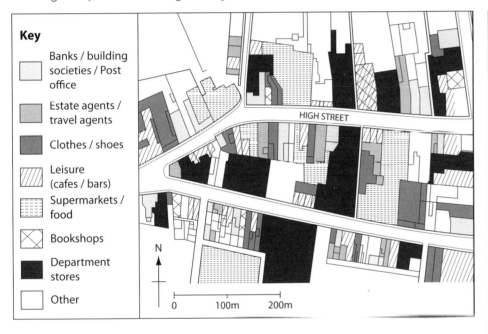

Key

- Banks / building societies / Post office
- Estate agents / travel agents
- Clothes / shoes
- Leisure (cafes / bars)
- Supermarkets / food
- Bookshops
- Department stores
- Other

HIGH STREET

N

0   100m   200m

OVER TO YOU

Urban investigations often include a land use survey as a common data collection method. Completing a second land use survey, perhaps identifying national / local business, or different functions, might provide additional useful data.

Use this space to plan your own land-use survey. Think about:

- Where you might get a base map?
- What types of information you might include?
- How you might carry out your survey?
- How you might present your data?

Land use transect

A land use transect is a survey along a line or road. It is a useful way to see how buildings and land use change with distance from a particular point. Land use along a transect can be shown on a map. However, the use of an annotated sketch or photograph will give a clearer impression of change. If you use a sketch or photographs, make sure the position of your transect is located on a map. The example below shows how land use changes as you move from the centre of a town in one particular direction.

Good Idea

When completing a land use transect always make sure you have some idea of the scale, including the relative size of the buildings. There may be a relationship between the width and height of buildings and the distance from a town centre.

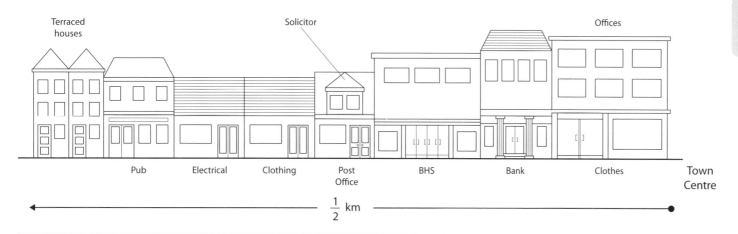

Terraced houses — Solicitor — Offices

Pub — Electrical — Clothing — Post Office — BHS — Bank — Clothes — Town Centre

$\frac{1}{2}$ km

OVER TO YOU

Fieldwork Focus question

Land use transects can be used in a variety of different investigations.

- Do you think one or more land use transects would be a useful source of information in your investigation? If so, explain why.

- For your investigation, explain how you might carry out a survey in order to produce a land use transect.
  Think about where your survey should start and end and how you might get some idea of scale.

Primary data: using cross-sections

Cross-sections are a useful way of showing physical landscapes. The following examples show how they can be used in different types of investigation.

A Coastal investigation – beach profiles

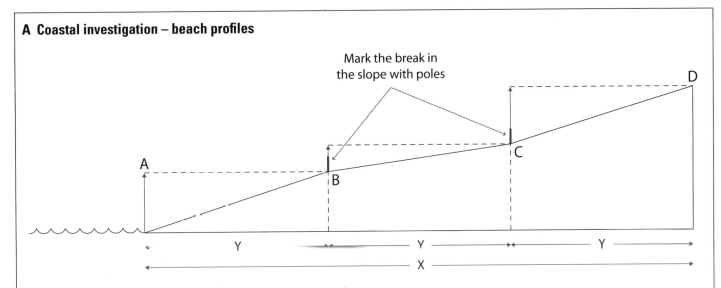

Mark the break in the slope with poles

1 Measure the width of the beach (**X**).
2 Mark the main breaks in slope and measure the distance between them (**Y**).
3 Using a clinometer (angle measurer) or by eye, work out the height at point A which would make it horizontal to point B.
4 Repeat this process up the slope to points C and D.
5 Present the data as a cross-section making sure that the vertical scale is not over exaggerated.

Good Idea

Use annotated photographs to help you describe and explain physical landscapes.

Develop this idea

Carrying out profiles at regular intervals along a beach (especially between groynes) may be a useful way of showing evidence of longshore drift.

Safety First

Coastal and River environments can be dangerous. Discuss with your teacher and parents what you are doing and do not work alone.

Be aware of particular sea or river conditions and watch out for:
- slippery and jagged rocks
- instability of cliff faces / slopes
- rapid change in waves / river flow
- changing tides.

B River Investigation – river cross-profiles

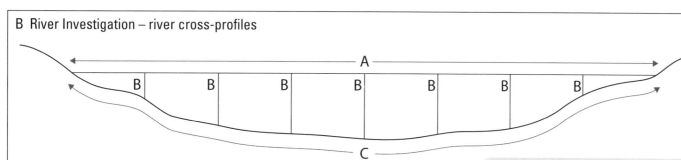

1 Measure the width of the river between the river banks (**A**).

2 Measure the depth of the river at regular intervals (**B**).

3 Present the data as a cross section making sure that the vertical scale is not over exaggerated.

4 For extra information you could measure the wetted perimeter (**C**).

Develop this idea

Carrying out cross-profiles at different points downstream will be a useful way of identifying changes to the river channel.

Safety First

Measuring the cross-profile of a river must be done using safety equipment and under proper supervision.

Only small streams / rivers should be used and never under flood conditions.

Research

A lot of useful information about coasts / rivers and general environments can be obtained from the Environmental Agency www.environmental-agency.gov.uk

Habitat / ecosystem investigations

Using a transect or cross-section is a useful way to show how plant life changes with slope (from the sea, a river, or simply across a slope). The process of measuring the slope can be done by using the technique for measuring a beach profile (page 52).

However, if the transect is over a long distance use an Ordnance Survey map to:

1 Identify the precise line of the transect.

2 Identify any breaks in the slope using the contour pattern and the scale of the map.

3 Work out the precise slope by using the contours of the map.

4 Walk along the line of the transect on the ground, identifying the different plants/habitats.

5 Present the data as an annotated cross-section making sure that the vertical scale is not over exaggerated.

Good Idea

This is a useful comination of both Primary evidence (data) and Secondary information (Ordnance Survey map). It could be developed further with the use of annotated photographs.

Primary data: using field sketches

Field sketches and photographs are an excellent way of showing visual information and are often very useful in physical geography investigations where there might be limited opportunities for the collection of primary data.

Field sketching

A field sketch gives you the opportunity to:

- identify important features and leave out things that are less significant
- show your interpretation of a place – how you see it
- add as much or as little detail as you like.

How to draw a field sketch

You do not need to be an artist to draw a field sketch if you follow a few simple rules.

1 Before you start always look carefully at the area you are going to sketch.

2 Decide the precise area you are going to sketch and stick to it!

3 Frame the area you want to sketch and divide it into a simple grid:

   a In a landscape sketch you might divide it into foreground, middle ground and background.

   b In an urban area which contains lots of buildings it might be easier to divide the area into grid squares.

4 Using the grid begin by adding the boldest features such as the horizon, roads, rivers and large buildings.

5 Add the detail which is important to your investigation.

6 Use shading or colour to show slope or structure. This will also help to highlight key features.

7 Complete your sketch by adding:

- a heading
- details about where it was drawn
- annotations identifying the key features.

Good Idea

Always locate the position of your field sketches or photographs on a base map. Use an arrow to show the direction you were looking when you drew the sketch or took the photograph.

Remember!

A field sketch is both a presentation skill and a means of identifying key points so it could be worth a lot of credit!

Good Idea

Why not base your field sketch on a photograph?

- Select a photograph.
- Draw a grid over the photograph to guide you.
- Do not include everything – pick out the most relevant points to your investigation.

The following example shows a field sketch in an investigation about river processes and features.

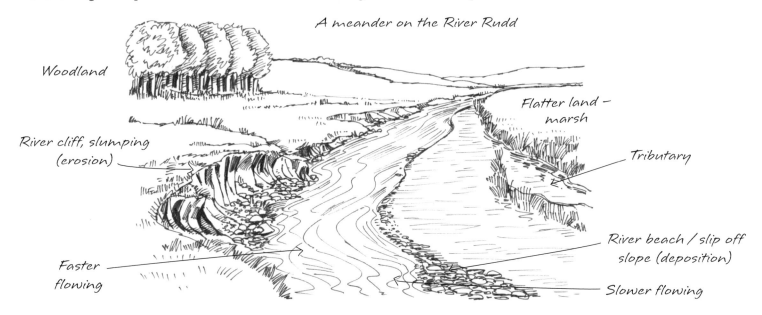

A meander on the River Rudd

Woodland

Flatter land – marsh

River cliff, slumping (erosion)

Tributary

River beach / slip off slope (deposition)

Faster flowing

Slower flowing

OVER TO YOU

Investigation question

1 How might field sketches be useful in your investigation?

2 What information/annotations might you include on a field sketch?

3 Use the following space to draw a practice sketch which might be useful to your investigation. Do not forget to add annotations.

Primary data: using photographs

Photographs are an excellent source of both primary and secondary data. However, when using photographs you need to consider the following points:

- Why is the photograph useful to my investigation?
- How can the photograph be used to make particular points?
- How am I going to show where the photograph was taken and the direction the camera was pointing?

Using photographs successfully

1 Don't use too many photographs – careful selection of appropriate photographs is important.

2 Each photograph you use must make a descriptive or analytical point.

3 Make sure you refer to your photographs within the written text.

4 Just like any other visual presentation – make sure each photograph has a clear title.

Oblique aerial photographs

Aerial photographs can sometimes be found on local websites.

They can often provide an excellent front cover illustration which clearly shows the nature and location of a study. The following photograph was used on the cover page of an investigation about coastal processes and features. It was also included within the investigation with detailed annotations.

Remember!

Photographs can also be secondary data if taken from another source of information.

Good Idea

A well chosen photograph on the front cover of your work will not only help to 'set the scene', it is also a presentation skill!

A word of warning!

It is easy to see photographs as a 'space filler' which then makes your completed work look more like a photograph album rather than a geographical report!

The important thing is to 'use' the photograph to make a point and to add annotations to develop the point.

Annotating photographs

The following examples show how annotated photographs can be used in both human and physical geography reports.

Traffic management town centre investigation

Major bank or building society

Older buildings – many with protection orders

National shops

Narrow pavements

Small pedestrianised areas around crossing point

A number of crossing points with traffic management

High Street (one way)

Parking on both sides – only narrow road area

Coastal management investigation

Large rocks protecting harbour entrance

High, fenced sea wall and promenade

Sloped sea wall to reduce the power of the breaking waves

Evidence of cliff slumping

Waves breaking over sea wall

Evidence of erosion

Old sea wall with metal edge

Tetrapods (large concrete 'jacks') protecting the sea walls

OVER TO YOU

Make a list of all the photographs that might be useful to your Controlled Assessment reports. Remember to state why each will be useful.
You could do this in table form.

Secondary data: using written information

Written information can come from a variety of sources, including:

- local newspaper articles
- business magazines
- Tourist Board information brochures
- local authority departments
- environmental management organisations (National Parks, Natural England, National Trust, etc.)
- specific environmental management areas (nature reserves, river parks, etc.)
- government departments such as the Environment Agency.

How can written information be useful?

Written information is a very useful source of secondary data. It can be used in a number of ways, including:

- helping to locate your Fieldwork Focus or Geographical Investigation
- helping to set the scene and give background information about the topic you are investigating
- providing facts and figures
- highlighting issues and conflicts
- identifying different opinions about a particular issue
- helping to describe and explain any management strategies that are linked to your topic of investigation.

How should you use articles?

Any written articles you find will not have been written specifically for your investigation! Consequently, it is important to identify the information that is relevant to your investigation.

This can be done in a number of ways, including:

1 Identifying the key points and making recorded notes in a table like the one shown below.

Source of information	Important points

Think!

Most government and private organisations have websites, which might have useful information.

Good Idea

Always make sure you have a record of all sources of information.

Never simply cut and paste written information into your report without any explanation of its importance.

2 Including the complete article and highlighting the key points. The following example shows an article used in an investigation about the Peak District National Park.

Headline expresses the pressure on the Peak District National Park.

Loved to death

Suggests an environment / economic conflict.

The Peak District National Park opened on April 17, 1951, and was rapidly followed by nine other parks. National parks are not just about environmentalism. They are pivotal in rural economies. On top of the annual government funding of around £26 million and the same again from EU and other sources, they attract 100 million visitors a year.

Massive number of visitors.

The Peak District park, for example, is the most visited park in the world after Mount Fuji in Japan. Each year its 550 square miles receive more than 30 million visitors.

Identifies key pressures.

Suggests weekend / seasonal pressures (honeypots?).

The pressures from vehicles and walkers intensify constantly, and the park is often close to breaking-point. On Sundays and bank holidays, traffic can be gridlocked around Longdendale.

Footpath erosion seen as a major problem.

Solving the traffic problems will not cure another great concern - erosion of paths by millions of walkers. When one stretch of the Pennine Way opened it was "a wee sheep track" but it grew to 60 metres in width. Now footpath teams work constantly to stabilise erosion.

Increasing pressure linked to building.

With the increase in visitor numbers there is increasing demand for new roads and the building of hotels.

© adapted from Carolyn Murrow-Brown, *The Times*, 14 April 2001.

Good Advice

This is a good way of:
- picking out the points that are closely linked to the question
- highlighting key issues / problems
- identifying different views / opinions.

OVER TO YOU

Use this table to list any secondary information sources you use.

Fieldwork focus investigation question: ....
....

Source of information	Important points
	• • •
	• • •
	• • •
	• • •

OVER TO YOU

Use this table to list any secondary information sources you use.

Geographical Investigation question: ....

Source of information	Important points
	• • •
	• • •
	• • •
	• • •
	• • •

Secondary data: using census information

What is a National Census?

A National Census is a population survey. In the United Kingdom a census is carried out every ten years, on the first year of each new decade: i.e. 1971, 1981, 1991, 2001, 2011 etc.

The National Census provides information at a range of scales, from neighbourhood areas containing a small number of people right up to national information about the whole country. The most useful scale for local Fieldwork Investigations is either neighbourhood or ward information (a ward is an area which usually contains 2000+ people).

However, comparing local area data with the national average can be very useful in some Geographical Investigations and census data is also very useful for identifying change over time.

What is included in the National Census?

The National Census collects information about a range of socio-economic characteristics, including:

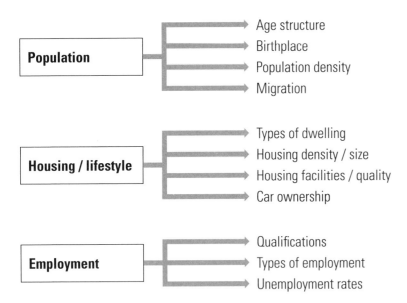

Population
- Age structure
- Birthplace
- Population density
- Migration

Housing / lifestyle
- Types of dwelling
- Housing density / size
- Housing facilities / quality
- Car ownership

Employment
- Qualifications
- Types of employment
- Unemployment rates

Find out more:

about the UK Census at the Office for National Statistics.

www.statistics.gov.uk/census

www.ons.gov.uk/census

The Census website also lists lots of other sources of information which might be useful.

Good Advice

Although the National Census is only carried out every ten years, local authorities collect information all the time. You can get a lot of background data, maps, photographs etc. via your local authority website. Each Local Authority has a number of departments, including planning, housing, transport, environment, etc.

– Would information from any of the local authority departments in your area be useful for your investigations?

Other sources of useful information:

The Index of Multiple Deprivation (IMD) uses seven sets of data to work out how deprived (poor) areas are. This can be used to compare different parts of urban areas. Find out more at: www.communities.gov.uk

Health information can be found at: Department of Health www.dh.gov.uk

How might you use census data?

The following examples suggest how census information is often a very useful source of secondary data to some investigations.

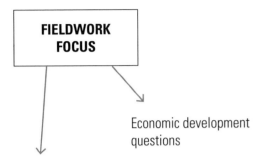

```
FIELDWORK
FOCUS
```

Economic development questions

Population / settlement questions

```
GEOGRAPHICAL
INVESTIGATIONS
```

Socio-economic census data might be helpful to a number of investigations based in the UK.

For example:

Disease - identifying rates / patterns of different types of disease.

- identifying why the patterns of disease varies (links to poverty etc).

Sport - Using sport to:
- develop run down areas (London Olympics)
- improve levels of health.

Crime - Identifying rates / patterns of crime.
- Considering the reasons why patterns of crime might vary.

Tourism - Thinking about population change / structure in declining tourism areas.
- Considering the importance of tourism to some areas.

OVER TO YOU		
Look at the types of data that are available on the sources mentioned on these two pages. Identify any information that might be useful to your investigations. Make a note of the source and type of information.	**Fieldwork Focus investigation**	**Geographical Investigation**

Secondary data: presenting census information

The following examples show some of the information that can be found in the National Census and different ways that it can be presented.

Good Idea

Remember to use a range of presentation techniques in your Fieldwork Focus investigation.

Line graph

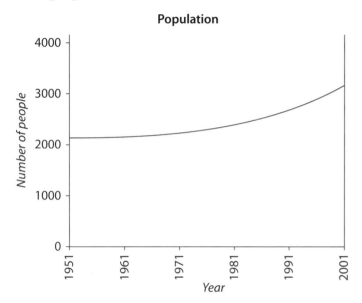

Multiple line graph

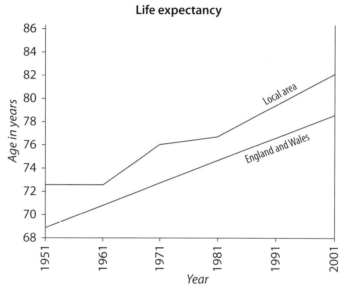

Bar graph

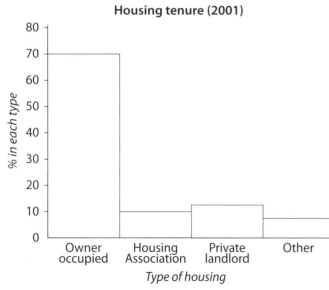

Multiple bar graph

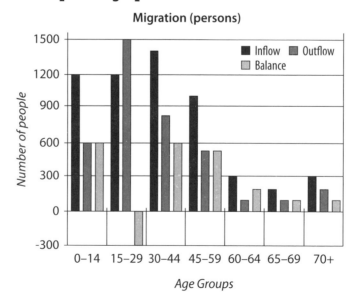

Good Idea

Multiple line and bar graphs are a useful technique for comparing information.

Pie chart

Housing type

Caravan / temporary dwelling

Detached

Flat / apartment

Terraced

Semi-detached

Divided bar graph

Housing type

Caravan / temporary dwelling

Detached	Semi-detached	Terraced	Flat / apartment	

0   100%

Choropleth map

A choropleth, or shading map can be used to show patterns within an area.

- The information being mapped needs to be divided into groups (classes). Four or five groups are usually needed to show a clear pattern without the map becoming too complicated.
- A colour or type of shading is used for each class, making sure the shading becomes darker for the higher values.

Population density

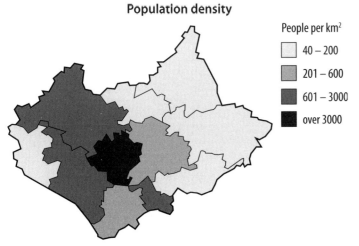

People per km²

- 40 – 200
- 201 – 600
- 601 – 3000
- over 3000

Table of figures

Employment (%)	
Agriculture / forestry	2.8
Fishing	1.4
Quarrying / mining	2.7
Manufacturing	14.2
Construction	6.8
Retail	18.3
Transport	6.5
Hotel / catering	16.3
Property management	7.8
Health / education / social services	23.2

Secondary data: using websites

Websites can be a useful source of secondary data for both the Fieldwork Focus investigation and the Geographical Investigation.

Fieldwork Focus

Although the Fieldwork Focus investigation is mainly based on primary data, secondary data can be used:

- to show a general understanding of the topic and provide background information
- as a source of local background information.

For example:

Local background

- Local authority website (planning, transport, services, housing, economic development etc).
- Local tourism offices.
- Many local businesses will have websites.

River / coastal studies

- Search key words such as erosion, deposition, landforms, management, flooding.
- The Environmental Agency website may be a useful starting point www.environment-agency.gov.uk .

Environmental issues

- Environmental management websites may be helpful, for example:

  National Trust – www.nationaltrust.org.uk

  English Nature – www.naturalengland.org.uk

  Forestry Commission – www.forestry.gov.uk

Remember!

For Level 3, the mark scheme says that you should have 'a wide range of sources, including fieldwork'.

Website information could be part of your 'wide range'.

OVER TO YOU

Use the following space for website planning for your Fieldwork Focus investigation.

1 Identify any websites that might be useful (including links).
2 Briefly describe the information that you might use and explain why it might be helpful.

Website (link)	Specific information	Why it might be helpful

Geographical Investigation

The Geographical Investigation is based on secondary research. Consequently, websites can be very useful in providing:

- background information about the theme of the investigation
- specific information linked to the investigation question
- different views and opinions about particular issues.

The following example shows how website research can provide a valuable source of information.

WEBSITE PLANNING SHEET	
THEME: *ENERGY*	Question: *Is wind power the energy for the future?*

General Information

Energy company websites — Shell – www.shell.co.uk ← *The Shell Foundation has a lot of information.*

— Shell Foundation – www.shellfoundation.org

— Shell (Renewables) – www.shell.com/renewables

— B.P. – www.bp.com

Energy management — www.renewableenergyworld.com

— www.energysavingtrust.org.uk

— www.energy.com

— *it might be useful to get a national perspective, for example: www.german-renewable-energy.com*

— *it might be useful to look at the issues associated with a wind energy development. Cefn Croes in Wales has a good range of information.*

Different views / opinions

Greenpeace — www.greenpeace.com

Friends of the Earth — www.foe.org

Think about
- Wind power (energy) in the UK
- Advantages / disadvantages of wind power.

Good Idea

- Focus on key words and locations in any web search.
- Media websites are always a good place to start, for example:
  www.bbc.co.uk
  www.telegraph.co.uk
  www.guardian.co.uk
  www.timesonline.co.uk
- The National Census may be a useful source of information for geographical investigations based in the UK.
  - www.statistics.gov.uk/census
  - www.ons.gov.uk/census

Remember!

It is important to show an understanding of different views / opinions in your Geographical Investigation.

OVER TO YOU

Use this website planning sheet to build up a list of useful websites and related ideas for your Geographical Investigation.

Theme	Question

Section 4: Putting your work together

Organising your Fieldwork Focus report

When planning your write-up remember that you are producing a report, not a book! You only have approximately 1200 words so that your final report needs to be well organised and logical.

Organisational structure

The following example will give you some ideas about how you might organise your fieldwork report.

COVER PAGE

- A well presented cover page helps to set the scene and shows that your work is well organised.

You must include:

- your name and examination number

- your school name and centre number

- the Unit code (B562) and title (fieldwork focus)

- the theme and question title for your investigation.

You could include: a well chosen photograph or map to highlight the topic being investigated.

INTRODUCTION - SETTING THE SCENE

- The first written section in most investigations is an introduction. This 'introduces' the question and gives a brief background to the investigation (including a map showing the location of the investigation).

DATA COLLECTION

- This is a short section which identifies and discusses the methods of data collection.

DATA PRESENTATION AND ANALYSIS

- This is where you present, describe and explain the data that you have collected.

CONCLUSION

- This is where you use the information that you have collected to answer the original question.

EVALUATION

- This is where you comment on how well your Fieldwork Investigation worked and how it could be improved and developed.

BIBLIOGRAPHY

- This is a list of the resources that you have used, including books, articles, websites etc (see page 80 for more detail).

General Advice

- Use clear headings to show the route of your investigation.

- Use annotations on data presentations (maps, graphs, photographs). This is a useful way of highlighting important points without using too many words.

- Make sure all maps, diagrams, graphs, photographs have clear titles.

- Use a 'report style' with clear titles, sub titles and descriptive information listed or put in boxes.

- Use blocks of written text for background description, explanation, conclusion and evaluation.

- Use a clear structure which is based around the mark scheme and the guidance given by OCR.

- If appropriate you could include a contents page with each page numbered.

Important guidance

You only have 1200 words for the Fieldwork Focus report so it is important that you make sure you say everything you need to without wasting words!

Each written section should be planned and written to make sure that:

- you have not missed out important points

- it is well organised and makes sense

- spelling, punctuation and grammar are accurate (remember the mark scheme! – pages 16–17).

Using the mark scheme to help you succeed: introduction

Setting the scene

- You should start by stating the question.

- **You should include:**
  - brief background information saying what the question is about
  - some understanding about the context of the question (what it is about in relation to your investigation)
  - a number of key questions which will help you break down the overall question being investigated
  - a clear description of the location of your investigation (a location map).

- **You could include:** a simple 'key words' or 'definitions' box where you identify and define the words that are important to your investigation. This is a useful technique because:
  - it is a useful way of showing understanding and knowledge
  - it allows you to refer to these words throughout your report without having to keep explaining them
  - it does not use up many words.

Mark scheme

Make sure you:

- describe the general topic of the investigation

- describe the topic in relation to your local area.

- use a variety of skills and techniques to locate and describe the study area.

OVER TO YOU
Fieldwork Focus question

Use this space to complete a rough draft of your written introduction.

Locating your investigation

Geography is always about place so it is important that you locate your Fieldwork Focus investigation effectively. Always think, 'If someone who does not know this area looked at my Investigation, would they be able to work out exactly where it was?'

Location can be 'described', but using maps is a much better option.

Why should you use maps to locate your investigation?

1 Using maps is a clearer and easier way of showing location.

2 Using maps is a presentation skill.

3 Using maps can save you a lot of words and fits in better with a 'report' style.

4 Maps can be annotated and used with photographs. This is a useful way of describing the most important points of your investigation.

The following examples show how two different methods have been used to describe the location of investigations. What are the strengths and weaknesses of each method?

Background - Investigation about traffic problems in Alton town centre.

The location of Alton

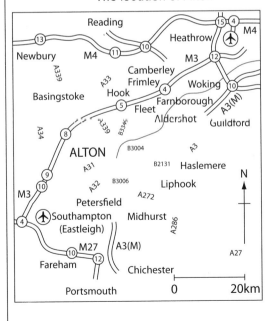

The main features of Alton town centre

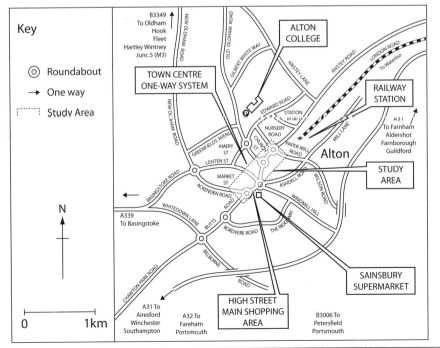

Do not forget

The mark scheme states that you should use a variety of skills to locate your investigation.

Good Advice

Published maps can be useful but can have too much detail. Remove what you don't need and highlight what is important to your investigation.

Using Ordnance Survey maps

Ordnance Survey (OS) maps can be an excellent source of information and can be used to produce both a general location map and a base map of the actual study area. Be careful to remove any information that you don't need and use annotations to highlight important points.

Examiner's comment

It is always useful to show the general area of the investigation and also a more detailed site map of the specific study area. In this example the site map might be better if it showed a smaller area surrounding the study area. It might then be possible to include a little more detailed information linked to the investigation.

Background - Investigation about land use change / regeneration in an urban area.

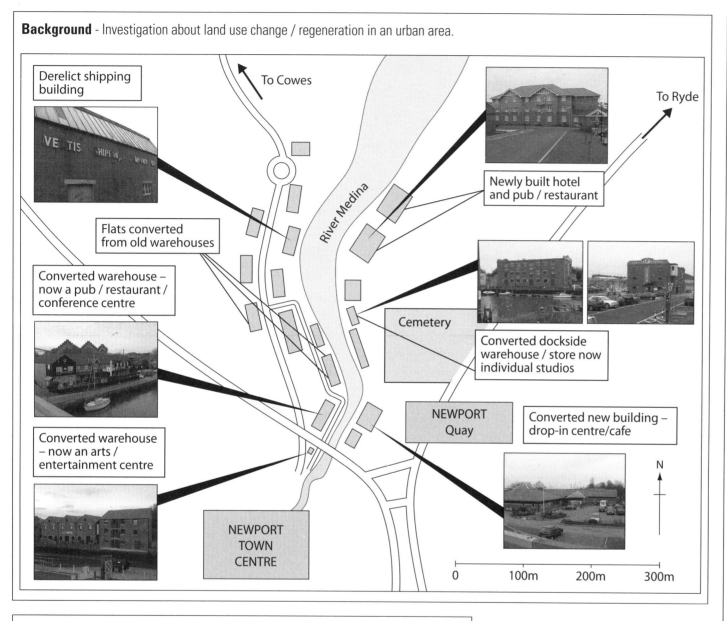

To Cowes

To Ryde

Derelict shipping building

River Medina

Newly built hotel and pub / restaurant

Flats converted from old warehouses

Converted warehouse – now a pub / restaurant / conference centre

Cemetery

Converted dockside warehouse / store now individual studios

NEWPORT Quay

Converted new building – drop-in centre/cafe

Converted warehouse – now an arts / entertainment centre

NEWPORT TOWN CENTRE

N

0   100m   200m   300m

OVER TO YOU
Fieldwork Focus question

List all the information that you need to include on the map showing your study area.

Examiner's comment

The map gives excellent detail about the study area and has very clear links to the topic being investigated. It uses a 'variety of skills and techniques' (mapwork, photographs, annotations). However, it does not have a title or a key. There is also a need for a general location map to show the actual location of the study area.

Using the mark scheme to help you succeed: data collection

You should include:

- a description and explanation of the data collection methods used
- some understanding of any problems or limitations with the data collection methods.

Useful techniques for the data collection section

Data collection table

A data collection table is a way of listing data collection methods and explaining why they were important. The following example shows part of a data collection table for an investigation about shopping patterns in a town centre.

Key questions	Method	Description	Importance of data	Problems / limitations	Possible development
Which is the busiest part of the town centre?	Pedestrian survey	Ten sites were chosen at 100m intervals across the town centre. People passing were counted for five minutes during a weekday (mid afternoon).	• Important to identify how busy different parts of the town centre are. • Important to see if there is a link between pedestrian numbers and location of car parks?	• It was unusually wet and cold so numbers may have been reduced. • Building work made counting in two locations difficult. • Narrow range of data.	• Data at different times of day / days of week may give a clearer picture. • Another four sites may give clearer results for the whole town centre.

Data collection map

If most of the primary data is collected in a relatively small area a data collection map is a useful way of showing the general location of the data collection points. The example shows a data collection map for an urban investigation about managing people and traffic management.

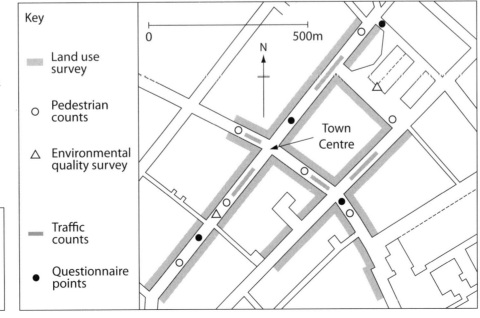

Key

░ Land use survey

○ Pedestrian counts

△ Environmental quality survey

━ Traffic counts

● Questionnaire points

0 ———— 500m

N

Town Centre

Examiner's comment

'A data collection map is also a presentation skill and could be used with annotated photographs to highlight important points'.

Examiner's comment

Data collection tables are an excellent way of showing the 'what, why and how' of data collection methods. However, to get the highest marks you must make sure that you include enough detail!

Good Advice

Always try to give some idea about sampling decisions i.e.:

- numbers of people
- types of people
- numbers of survey sites
- location of survey sites.

Explanation of questionnaires

Most investigations use questionnaires as a source of primary data. A questionnaire summary, like the one below, can be used to explain the importance of the questionnaire to the investigation.

This question gives information about the stability of the population and rate of change. →

People who travel outside the town may rely less on local services. →

People who do not have a car may need more local transport services. →

This will give a general impression about people's views. →

This will give an idea about what people want. It is a mixture of shops and social services – the importance of each can be judged using this information. →

INVESTIGATION ABOUT THE SERVICES IN A LOCAL TOWN

1 How long have you lived in the town?
☐ 0–5 years   ☐ 6–10 years
☐ 11–15 years ☐ 16+ years

2 Do you travel to work outside the town each day?
☐ Yes   ☐ No

3 Do you own a car?
☐ Yes   ☐ No

4 Do you feel that the local services in the town are adequate?
☐ Yes   ☐ No

5 Which of the following services would you like to see in the town?

Leisure centre ☐   Improved parking facilities ☐
Hospital ☐   Retail park ☐
Cinema ☐
College ☐
Others (please list) ....

Make sure - that when carrying out questionnaires you always explain the sample size (the number of questionnaires) and how / why you selected the particular respondents.

Good Advice

If you have used a questionnaire include one copy and explain why each question is important to your investigation.

OVER TO YOU
Fieldwork Focus question

Use this space to make notes about the information you need to include in your data collection section. (You could use a table.)

When you have finished: - check that everything important is included
- use your notes to write up your data collection section.

Using the mark scheme to help you succeed: data presentation and analysis

You should - use a range of presentation methods, including more complex techniques

- make sure that all data presentations are neat, accurate and correctly labelled
- identify the key points from the data – those that are most relevant to the original question
- explain what each data set shows and identify any important links between data sets.

Presentation

Presentation is about:

- the visual methods used to present information (maps, diagrams, graphs, photographs etc.)
- using methods that are appropriate, complete and accurate.

Building up a range of presentation methods

Presentation methods can be used throughout your report. The following examples are just a few of the possibilities

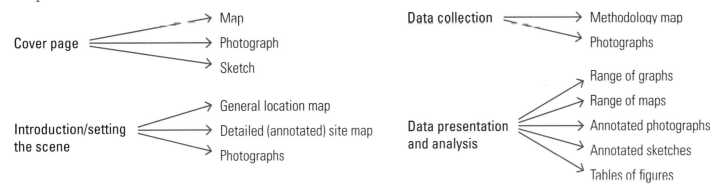

Mark scheme

Make sure you:

- present your data using an appropriate range of maps, graphs and diagrams
- thoroughly describe and explain your evidence
- identify and analyse the most important information so that it can be used to make detailed conclusions
- produce clear written work with accurate spelling, punctuation and grammar.

Good Ideas

- Use the 'collecting and presenting information' section to get ideas for data presentation methods. (pages 28–67)
- Look through textbooks to get ideas about presentation methods.

OVER TO YOU
Fieldwork Focus question

Use this space to create a checklist of the presentation skills you could use throughout your report.

Analysis

Analysis is where the collected data is explained and linked back to the original question. The key parts of data analysis are:

- describing each data set (what does it show?)
- explaining or suggesting reasons (why does it show this?)
- drawing out important links between the data sets
- using the most significant evidence from the data to reach a conclusion which answers the original question.

Useful data analysis techniques

Using tables

Tables are a useful way of summarising questionnaire data. The following examples show how they can also be used to identify the key points in relation to a question or issue.

> **Using statistical techniques**
>
> Data is often collected in the form of numbers (statistics). There are a number of different techniques that can be used to describe and explain statistics. Some of these can be seen on pages 91–93.

Investigation about pedestrianisation in a town centre.

Advantages of pedestrianisation	Disadvantage of pedestrianisation
Will make the area safer	Will cost a lot of money
Improve the shopping environment	Disruption while being put in place

Investigation considering different methods of traffic management.

Method	+	−
One way system	Faster traffic flow...	Confusing Requires road changes
Increase number of traffic lights	Slows traffic so: safer fewer accidents	May increase congestion

OVER TO YOU
Fieldwork Focus question

Use this space to identify the main points from your data. Remember to consider how important these points are in relation to your original question.

Using the mark scheme to help you succeed: conclusion

You should:

- start by re-stating the question
- answer the original question by using the strongest evidence from your data collection
- if there is not an obvious answer, make sure you write a balanced conclusion which uses evidence from both side of any discussion.

OVER TO YOU
Fieldwork Focus question

Use this space to consider the main points of your conclusion and identify evidence that you could use to support those points.

Mark scheme

Make sure you:

- use the interpretation of evidence to reach a thorough and detailed conclusion
- produce clear written work with accurate spelling, punctuation and grammar.

Examiner's comment

The most successful investigations identify the main points from the data and suggest clear reasons for the results of the data collection. They then identify the most important evidence and use this to return to the original question and reach a detailed conclusion – backed up by strong evidence!

Evaluation

The evaluation pathway

The mark scheme identifies three key elements to the evaluation of your investigation. It might be helpful to use your evaluation to make sure that each of these elements is considered. However, make sure you explain important links between the three elements. The following diagram may provide a useful starting point to think about your evaluation.

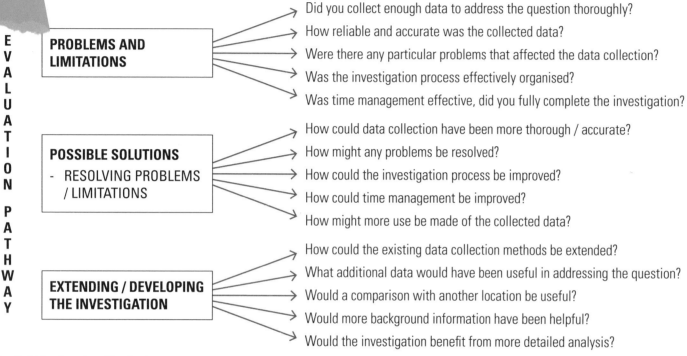

EVALUATION PATHWAY

PROBLEMS AND LIMITATIONS
- Did you collect enough data to address the question thoroughly?
- How reliable and accurate was the collected data?
- Were there any particular problems that affected the data collection?
- Was the investigation process effectively organised?
- Was time management effective, did you fully complete the investigation?

POSSIBLE SOLUTIONS
- RESOLVING PROBLEMS / LIMITATIONS
- How could data collection have been more thorough / accurate?
- How might any problems be resolved?
- How could the investigation process be improved?
- How could time management be improved?
- How might more use be made of the collected data?

EXTENDING / DEVELOPING THE INVESTIGATION
- How could the existing data collection methods be extended?
- What additional data would have been useful in addressing the question?
- Would a comparison with another location be useful?
- Would more background information have been helpful?
- Would the investigation benefit from more detailed analysis?

Useful Evaluation Techniques

The following two examples show techniques that can help to identify important points which can then be used as part of your evaluation.

Commenting on each data set

An evaluation table (go back to your data collection table if you have used one - page 72).

Method	Problems / limitations	Solutions	Development
Pedestrian flow survey	• Poor weather may have given unrealistic results. • Some people may have been missed. • Only completed on one day for five minutes.	• Complete during different weather conditions. • Have two people completing each count. • Complete on different days at different times for greater accuracy.	• A greater number of counts at more places on different days at different times. • Assessing types of people (male / female, etc).

Commenting on the whole process

- Identifying problems and solutions

Problems	Solutions
• Only collecting data on a weekday may have affected the overall pattern of results.	• Collecting data at the weekend would give a more reliable impression.
• Poor weather on the day of the data collection may have affected traffic / people numbers.	• Collecting the same data on a sunny day might show if the weather makes any difference.

- Identifying strengths and weaknesses

Strengths	Weaknesses
• Good range of traffic flow data.	• Base map not up to date.
• Photographs clearly identify the key issues.	• Limited number of questionnaire respondents.

Remember!

Data can be objective (facts / figures) and subjective (opinions). Subjectivity may affect the reliability of data.

OVER TO YOU

Fieldwork focus question

Use this space to consider the main points for your evaluation. Use the completed grid as background notes to help you produce your final evaluation.

PROBLEMS / LIMITATIONS	•
	•
	•
	•
POSSIBLE SOLUTIONS TO PROBLEMS / LIMITATIONS	•
	•
	•
	•
EXTENDING / DEVELOPING YOUR INVESTIGATION	•
	•
	•
	•

Organising your Geographical Investigation report

Remember, you are producing a short, structured report. You only have approximately 800 words so your report needs to follow a clear plan and be well organised.

Organisational structure

The following example will give you some ideas about how you might organise your Geographical Investigation report.

COVER PAGE

- A well presented cover page helps to set the scene and shows that your work is well organised.

You must include:

- your name and examination number
- your school name and centre number
- the Unit code (B562) and title (Geographical Investigation)
- the theme and question title for your investigation.

You could include: a well chosen photograph or map to highlight the topic being investigated.

LOCATION AND BACKGROUND INFORMATION

- This is an introduction which gives background information about the topic and makes the locational context clear.

INVESTIGATION QUESTION

- This is where key ideas from the question are considered. The question can then be broken down into a number of smaller key questions.

ISSUES ARISING OUT OF RESEARCH

- This is where your research is used to show that the question raises certain issues which people have different views / opinions about..

CONCLUSION

- This is where you used your evidence to answer the original question and make observations about future situations in relation to the question.

BIBLIOGRAPHY

- This is a list of resources that you have used, including books, articles, websites and so on. You should aim to include the author of the source (where relevant), the title of the source, and the date the source was created. An example is given on the next page.

Important guidance

You only have 800 words for this report so make sure you say everything you need without wasting words!

Plan and write each section in rough and then check to make sure that:

- you have not missed out important points
- it is well organised and makes sense
- spelling, punctuation and grammar are accurate (see the mark scheme on pages 26–27).

Good Advice

- Use clear headings to show the route of your investigation and use a clear structure based on the OCR mark scheme and guidance.
- Use annotations on data presentations (maps, graphs, photographs) to highlight important points without using too many words.
- Make sure all maps, diagrams, graphs, photographs have clear titles.
- Use a 'report style' with clear titles, sub titles and descriptive information listed or put in boxes.
- Use different presentation techniques to convey information and make points. This will save words and make your final work appear more 'report style'.

Author	Title	Web address	Publisher	Page	Date

Completing your investigation report

When writing up your investigation report a useful first step might be to break the question down by making notes which relate to the main demands of the mark scheme (pages 26–27). These notes can then be used as a guide and checklist when writing up your final report. The following examples show how this could be done.

Investigation question *Should the UK develop wind power in the future?*

Locational context

- Based on UK energy supply
- Could have a broader context (where UK gets energy resources from – oil / gas etc.)

Background questions / ideas

- What is the energy demand in the UK?
- How is energy demand currently satisfied?
- How long are existing energy resources likely to last?
- Will renewable energy be able to satisfy rising demand?
- What is the historical context of wind energy in the UK?
- Could wind energy fill the future 'energy gap'?

What are the issues

- Impact on local enviroments
- Impact on local economies
- Sustainability of wind power
- Relative cost of electricity produced by wind power

Information / data

- Timeline of wind energy use / development in the UK
- Global situation of wind power development
- Current likely energy supply / demand
- How UK energy mix has changed
- Likely decline of oil / coal / gas
- Likely growth of renewables
- Relative cost of electricity from different sources

Different views / opinions

- Energy companies
- Electricity supply companies
- Confederation of British Industry (CBI)
- Government
- Environment groups
  - Greenpeace
  - Friends of the Earth

> **Good Advice**
>
> You could use a questionnaire to get an understanding of local knowledge / opinions.

Investigation question *How are growing visitor numbers putting pressure on a National Park you have studied?*

Locational context

- Must be **one** specific National Park
- Could be in any county that has National Parks

Background questions / ideas

- Where is the National Park?
- What are the characteristics of the area?
- How many people visit the National Park?
- What is the pattern of visitor numbers,
  - across the area?
  - throughout the year?
- What are the effects of visitors to the National Park?
- What enviromental pressures are caused by visitor growth?
- What socio-economic pressures are caused by visitor growth?
- How are these pressures likely to change in the future?

Information / data

- Major attractions of the area
- Visitor numbers
- Economic / Environmental background
- Description of the existing pressures,
  - rates / costs of erosion of footpaths
  - number of second homes
  - car numbers in busy periods.

Different views / opinions about the growth of visitor numbers

- Local people ⟶ who rely on tourism
- Local business ⟶ who do not rely on tourism
- National Park managers
- Environmentalists
- Visitors – different user groups

OVER TO YOU
Investigation question

Use this space to break your question down in a similar way to the previous examples.

Using the mark scheme to help you succeed: location and background information

You should start by stating the question.

You should include:

- background information saying what the question is about
- some understanding of the locational context of the question (the place(s) that are important to the question)

Useful techniques

Annotated maps

Locational context is best shown by using an annotated map. This could also include useful background information.

Example: Investigation question – 'How are growing visitor numbers putting pressure on a National Park you have studied?'

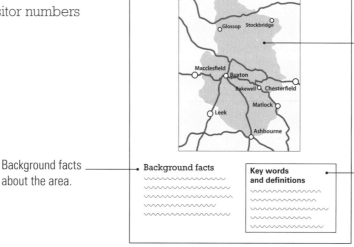

Location and background information

Background facts about the area.

Map of Peak District National Park with key features included (main roads, settlements, places of interest, attractions etc.).

Key words / definition box.

Mark scheme

Make sure you:
- show an understanding of the locational context of the question
- show an understanding of the geographical background to the question.

Key words

You could include a 'key words' or 'definition' box where you identify the key words.

Example: Investigation question - 'Should the UK develop wind power in the future', important key words might include:

- finite resources
- renewables / non renewable
- energy mix.

This is a useful technique because:

- it shows background understanding and knowledge
- it allows you to refer to these words in your report without having to keep explaining them
- it does not use up many words.

OVER TO YOU
Investigation question

Use this space to plan the opening section of your investigation report.

Start by asking yourself 'What is the question about'.

List the important 'key words' you might want to use.

WORD	DEFINITION
•	•
•	•
•	•
•	•
•	•

Using the mark scheme to help you succeed: investigation question

You should start by stating the question.

You should include:

- a brief description of the types of evidence required with some appreciation of why it is important.

You should have an understanding that the investigation question might raise other questions which need to be considered.

Useful techniques

- Considering a question often means starting by thinking about other, related questions. A diagram, such as a spider diagram or flow chart might be a useful way of showing how ideas are linked.

Example:

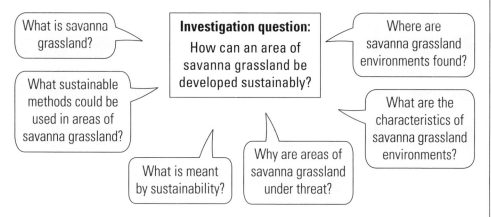

- A simple table could be used to list and explain the importance of evidence being used to address the investigation question.

Mark scheme

Make sure you:

- show an awareness of the key ideas raised by the question
- consider the range of appropriate evidence you might use to address the question.

Good Advice

Use the 'collecting and presenting information' sections to get ideas about gathering evidence. (pages 28–67)

OVER TO YOU
Investigation question

1 Use this space to identify important questions that are clearly linked to your original question.

2 Complete the table below by noting the evidence that you are going to use to answer the question. Briefly explain why the evidence is important.

Evidence	Why it is important
•	
•	
•	
•	
•	
•	
•	

Using the mark scheme to help you succeed: issues arising out of research

You should include:

- any background knowledge / case studies that you have gathered
- views of different people / groups about issues raised by your research.

Useful techniques

- Questions will often require background information or case study material, which might include quite a lot of information. This can be presented using a variety of visual techniques. This saves you using large blocks of text and might also help to show an understanding of the locational context of the question.

Example: Question about savanna grassland ecosystems.

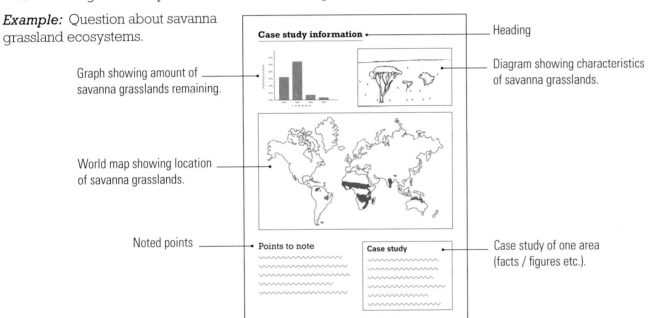

Graph showing amount of savanna grasslands remaining.

World map showing location of savanna grasslands.

Noted points

Case study information • — Heading

Diagram showing characteristics of savanna grasslands.

Case study of one area (facts / figures etc.).

Points to note

Case study

- Using speech bubbles is a technique often used to express different views / opinions. Make sure you identify key points and say who is actually expressing the opinion!

Example:

We need to cut down the trees for firewood in order to grow food.

Local subsistence farmer

Commercial farming is big business and creates lots of jobs.

Commercial farmer

Removing the vegetation means loss of animal habitats and will increase the risk of desertification.

Environmentalist

Mark scheme

Make sure you:

- select appropriate evidence from a range of sources
- apply knowledge of different opinions / attitudes about related issues
- explain why people might have different opinions/attitudes.

Remember!

Your Geographical Investigation must include secondary research.

Good Advice

- Use the 'collecting and presenting information' sections to get ideas for data presentation methods. (pages 28–67)
- Look through textbooks to get ideas about presentation methods.

OVER TO YOU

Investigation question

Use this space to make a note of the background information/case studies you might use and the different views and opinions raised by your research.

Background Information / case studies

Different views and opinions

Using the mark scheme to help you succeed: evaluation of research

You should include:

- a list of the evidence used with comments about its reliability

Points you might consider (there may be others!)

- How old is the information / data?

- Is the information accurate?

- Are the views and opinions objective (facts) or subjective (opinions)?

- Do you have enough information to make a sound judgement?

- What other information might have been useful?

- How reliable is any estimated data?

- Can facts and opinions be easily checked?

- Is there any bias in the information?

OVER TO YOU
Investigation question

Use this space to note down the main points of your evaluation.

Your final evaluation could be done in a table where you look at each set of data separately or as a paragraph where you evaluate the whole process.

Using the mark scheme to help you succeed: conclusion

You should:

- include a description and explanation of your collected evidence
- identify and use the strongest research evidence to support your conclusions
- show an understanding of how decisions taken about particular issues might affect future situations, for example:

Investigation question — How are growing visiter numbers putting pressure on a National Park you have studied?

Ideas could include

— enviromental pressures

— social pressures

— economic pressures

Growing visitor numbers create both advantages and disadvantages. Encouraging visitors puts pressure on the enviroment and could damage the very landscape that people come to see. This might mean the area is less attractive so the number of visitors dicreases, harming the local economy.

At the same time the growing number of visitors creates demand for second home ownership, pushing local house prices up. In the long term this could mean local people cannot afford to buy houses and have to leave the area. This could lead to the closure of local services like primary schools and shops.

Mark scheme

Make sure you:

- analyse and interpret information to reach a conclusion
- apply ideas from the collected evidence in your conclusion
- show awareness about how decisions about issues could affect future sustainability

Using statistical techniques

Research data is sometimes collected in the form of numbers (statistics) which can provide useful evidence when writing your conclusion.

There are a number of different techniques that can be used to describe and explain statistics. Some of these can be seen on pages 91–93.

OVER TO YOU
Investigation question

Use this space to plan your conclusion. Think about:

1 The most significant factors from your research.

2 The main points that you want to make in your conclusion (identify evidence to support these points).

3 How decisions about the issues identified might affect future sustainability (show an understanding of what is meant by sustainability in relation to your study topic).

Section 5: using statistical techniques to describe and explain data

Describing data

Data is often collected in the form of numbers (statistics). There are a number of simple ways of describing statistics, including:

- calculating averages
- calculating the median (the middle value of ranked data)
- calculating the mode (the most frequently occurring number)
- describing the minimum and maximum numbers
- calculating the statistical range (largest minus smallest)
- drawing a dispersion diagram.

Example of a dispersion diagram

Question: 'How often do you visit the following shopping areas each month?'

Each dot represents one person.

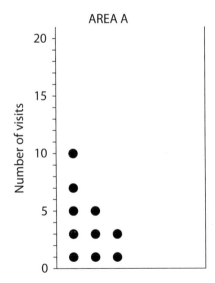
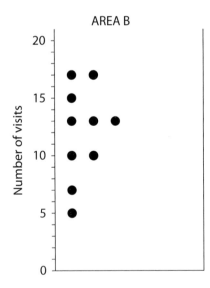

Using statistics to explain / analyse relationships between data

The following example shows how statistical techniques can be used to examine the relationship between two sets of data.

Using statistical techniques: an example of a shopping investigation

Aim: To see if the number of pedestrians decreases with the distance from a town centre.

Data collection method: Ten locations were chosen at varying distances from the town centre. At each location the number of pedestrians was counted over a five-minute period.

Collected data for shopping investigation.

Location	1	2	3	4	5	6	7	8	9	10
Distance from town centre (m)	10	50	100	80	70	200	150	400	300	250
Number of pedestrians (nearest 10)	320	280	270	260	290	200	210	90	190	170

Presenting the data as a scattergraph

To draw a scattergraph:

1 Draw and label the two axes.

2 Choose scales to cover the range of data.

3 Plot the data using dots.

4 Put on a line of best fit (do not just join up the dots!).

What does a scattergraph show?

- **Positive relationship** – as one data set increases, so does the other.
- **Negative relationship** – as one data set increases, the other decreases.
- **No relationship** – no real pattern is evident, so the relationship is unproven.

What does the scattergraph of the shopping investigation data suggest?

1 There is a negative relationship – the number of pedestrians falls as distance from the town centre increases.

2 It is quite a strong relationship since all the dots are close to the line of best fit.

3 Some dots do not quite fit the pattern (some explanation for this would be useful).

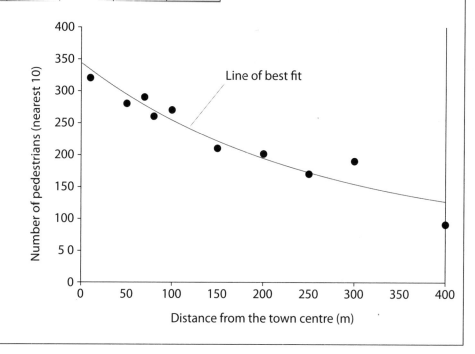

Testing the strength of a relationship

Looking at the scattergraph of the shopping investigation data on the previous page, it is clear that the two sets of data are related. The strength of a relationship between two sets of data is called a correlation. This can be calculated using the Spearman Rank Correlation Coefficient (RS). This is a statistical calculation which always gives a result from −1 (negative correlation) to +1 (positive correlation); the nearer to 1 the stronger the link between the data.

How do you calculate RS?

1 Rank both sets of data from the highest to the lowest.

Distance from town centre	Rank	Number of pedestrians	Rank	Difference between ranks (d)	d^2
400	1	90	10	9	81
300	2	190	8	6	36
250	3	170	9	6	36
200	4	200	7	3	9
150	5	210	6	1	1
100	6	270	4	2	4
80	7	260	5	2	4
70	8	290	2	6	36
50	9	280	3	6	36
10	10	320	1	9	81
					Total (Σ) = 324

2 Use the formula:

$$RS = 1 - \frac{6 \times \Sigma d^2}{n^3 - n}$$

Where:

n = number of observations

d = difference between ranks

Σ = total d^2

3 Using the example above:

$$RS = 1 - \frac{6 \times 324}{1000 - 10}$$

$$= 1 - \frac{1944}{990}$$

$$= 1 - 1.196$$

$$= -0.96$$

This shows that there is a very strong negative relationship between the number of pedestrians and the distance from the town centre. In other words as you move away from the town centre the number of pedestrians decreases.